# Sailing at Night

# Sailing at Night

### THE SEAMANSHIP SERIES

## Richard Henderson

International Marine
Publishing Company
Camden, Maine 04843

Photographs on pages 15 and 28 are reprinted courtesy of Tom Hindman.

©1987 by International Marine Publishing Company

Typeset by The Key Word, Inc., Belchertown, Massachusetts
Printed and bound by BookCrafters, Inc., Chelsea, Michigan

Published by International Marine Publishing Company
21 Elm Street, Camden, Maine 04843
(207) 236-4342

**Library of Congress Cataloging in Publication Data**

Henderson, Richard, 1924–
    Sailing at night.

    (The Seamanship series)
    Includes index.
    1. Sailing.  2. Navigation.  3. Sailing—Safety measures.
I. Title.  II. Series.
GV811.5.H466  1987          797.1'24          86-20109
ISBN 0-87742-226-5

*Dedicated to my first cousin and
old friend Ed Henderson,
with whom I've enjoyed
many an overnight race.*

# Contents

# 1

# **Prerequisites to Sailing at Night**

There are few experiences so satisfying as sailing at night. The boat seems to rush into the darkness, leaping ahead in a fresh breeze or easily ghosting in a calm, appearing to sail much more swiftly than she actually is. The diminishing of sight is compensated by greater awareness through the other senses. Sounds are especially vivid: the sloush or slap of the bow wave, hiss of the wake, leech flutter of the sails, and soft sighing of wind in the rigging. What little we can see are visions to be savored: moonlight bouncing on the water, spray colored by the running lights, phosphorescence sparkling in the sea, and even the glow from a shipmate's pipe. Yet still greater pleasures of sailing at night come from the feeling of accomplishment and a slightly defiant sense of freedom.

Night sailing, of course, is not always done for gratification alone. Very often it may be the necessary or sensible thing to do. On long passages, obviously, sailing after dark is unavoidable. Even on fairly short passages it may make sense to leave your home port or other familiar harbor in

1

the dark so that you can be assured of arriving at a less familiar port in broad daylight or so that you can leave on a favorable tide. In poorly marked tropical waters with coral reefs, it is often the safest policy to arrive not long after midday, when the shoals are most visible.

On the other hand, there are certain occasions when difficult landfalls may be desirable after dark or at least in the twilight, when lights from navigational aids and the loom over land can be seen. Then too, night sailing may prove necessary when a usually short haul becomes unexpectedly long due to foul currents, rough seas, head winds, fog, engine breakdown, or some other reason. If you're a serious cruising sailor, you should figure that sooner or later you'll have to operate your boat after dark.

## PREPARING THE BOAT FOR NIGHT SAILING

Boat preparation for night sailing begins with installing fittings and laying out gear in a manner that minimizes the risk of tripping crew or fouling lines and sails. Keep in mind that Murphy's Law, predicting that things will go wrong when they can, is particularly evident at night.

Every effort should be made to keep the decks clear, and if necessary gear is a potential obstruction, it should be marked with reflective or white tape or else painted a color that contrasts with its surroundings. Care must be taken to coil and stow loose lines, and you should never allow sails to lie spread out on the deck. Keep the hatches closed whenever there is a possibility of someone stepping into them. Other booby traps are loose seat cushions that can slide underfoot and cause tripping. On boats of any size there should be lifelines and pulpits. Install ample handholds and skidproof slippery surfaces (more about this in Chapter 5).

Do your best to improve visibility. Lower spray dodgers and awnings, and if possible, avoid carrying a dinghy on the cabintop. See that any hinged hatch cover is folded down.

*This boat's genoa jib is a good one for night sailing in that the foot is sufficiently high for visibility and reasonable clearance of lifelines.*

It is best to carry a jib with a high clew and a sufficiently short luff that a modest tack pendant can be rigged to raise the foot, allowing the helmsman to see under it. When there is a gooseneck track, the main boom should be up as high as it will go.

Essential equipment should be put in a handy place so that it will not have to be looked for when needed. To avoid fouling coiled lines, turn the coils so that the line runs off the top. It's a good idea to run an antifouling line from the forward side of the mast, about seven feet up, to an eye on deck about three feet from the mast base. This will keep an overlapping jib from fouling on mast winches or cleats when you're changing tacks. Check that sheets cannot catch on anchors, spinnaker poles, boat hooks, or other equipment stowed on deck. Spinnaker pole chocks should have rounded corners, and clam cleats should have spring gates to ensure against fouling. Try to anticipate what could possibly snag or tangle.

Another aspect of boat preparation is soundproofing. Be sure there is plenty of noise-inhibiting insulation in the engine room and under the engine hatch, because it is important that you be able to hear sound signals, navigation aids, and shouts from other boats and from your own bow lookout, especially when powering through populated waters in the dark. Try to damp the noise from vibrations, and tie off your halyards so they don't beat the mast. Radios blaring music are a bad idea. Aside from the safety of quietness, it is ever so pleasant and helpful to sailing when you can hear the sound of the water.

## SEEING AT NIGHT

Before starting to sail at night, it is helpful to learn something about night vision, as a little knowledge of this subject can enable you to improve your vision in the dark and semidarkness. One important principle to understand is the value of peripheral vision. Light receptors in the eye's retina consist of neurons called cones and rods. The former are tightly packed at the very center of the retina (the inner back part of the eyeball on which light rays are focused) and effectively handle daytime or well-lighted conditions. This central area of the retina, called the fovea

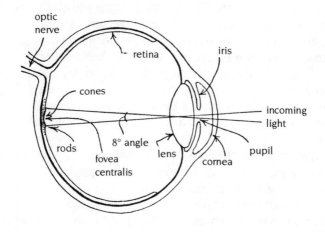

*The human eye.*

centralis, provides best vision when you look directly at a viewed object.

Rods take care of vision at night or in very dim light. Their location slightly away from the center of the retina, surrounding the fovea spot, means that rod vision works best if the dim light falls on the retina about four degrees off-center. This condition is met when a person looks slightly to one side or above or below the viewed object rather than directly at it. If you've caught a brief glimpse of a channel marker or crab pot and then lost sight of it, look not directly at the spot where you saw it but slightly to one side, and the object will most likely reappear. When the object lies on the horizon, try looking a little above or below. The technique of using peripheral vision can be learned through practice. It has been taught effectively to fighter pilots, ship lookouts, and others who must "see" in the dark.

It is important to keep your eyes moving and avoid lengthy staring, which can impair visual perception and cause fatigue. Blinking and other eye movements help restore eye efficiency. Lookouts on ships are often taught to search the waters in a systematic rectangular pattern to keep their eyes moving as well as to ensure that all areas are covered. A method that seems to help me when I'm looking for an object at night is to move my head in a circular manner. This keeps my eyes moving even if they are focused on a small area, and it assures me that an unsuspected piece of gear or rigging is not blocking my view.

Warnings are often issued against pressing binoculars firmly against the eye sockets for prolonged periods, as this can cause minor damage and impair vision. One also hears about the so-called blind spot where the optic nerve attaches to the retina, but this won't cause any problem as long as you keep both eyes open simultaneously. It should be obvious—but all too often isn't—that a navigation aid or other object is easier to find in the dark if you know in advance what you are looking for. Partly for this reason, you should always carry the *Light Lists*, which provide descriptions of all navigation aids. These lists, as well as other matters having to do with navigation, will be more fully discussed in Chapter 4.

What can be seen at night—for example, the dark silhouette of a shoreline against a lighter sky— will usually appear closer than it actually is. This may be partly due to the fact that the shore's dark reflection blends with its silhouette, imparting the appearance of greater vertical relief. Navigation aids, too, may seem closer than they really are. Indeed, racers have been fooled into changing sails too early when approaching a turning mark at night, thinking they were just about to round it. The observer's height, as well as the marker's height, can affect visibility. An unlighted buoy may become more visible if you can get low enough to see its top above the horizon when the sky is lighter than the sea or shoreline. On the other hand, a lighted buoy is usually more easily seen when the observer is high, particularly if the light is so far away that it appears to be dipping below the horizon. The *Light Lists* are valuable in describing the visibility range of lights, but remember that atmospheric conditions, particularly a bank of fog, can reduce the range or even completely obscure the light.

According to an early theory (Fechner's Psychophysical Law, which relates color brightness to the amplitude of light waves), red and yellow disappear first and blue fades last in diminishing light. On the other hand, eye doctors I consulted say that visibility in dim light is mostly a matter of contrast, and that white (a mixture of all colors) against a dark background provides maximum visibility. This apparent contradiction may be at least partly due to the fact that the cones see color and are still in operation in dim light, but the rods, which see no color, gradually take over as the light fades into darkness. For greatest visibility you want the brightest color with plenty of contrast in dim light, but in darker conditions you want the greatest possible contrast regardless of color.

Although white beside black provides maximum contrast in color value, you have to be careful about this combination on the water, because white blends nicely with foam. Beware of white life rings, for example. An eye doctor/sailor friend of mine experimentally threw one overboard on a dusty night when whitecaps were everywhere, and he never could find the buoy. Yellow is a high-intensity color offering almost as much contrast as white; for this reason I prefer it for life rings, foul-weather gear, and personal flotation devices.

The human eye takes a considerable time to adapt fully to the dark after it has been exposed to bright light. One hears various figures, but the eye doctors I consulted suggested 12 to 40 minutes as an average, depending, among other things, on the brightness of light to which the eye was exposed. A person's age seems to have no significant effect on adaptation time, but individual differences in body chemistry do. According to a prevalent theory, some individuals are lacking in rhodopsin, a substance chemically related to Vitamin A and found inside the rods, and this deficiency results in night-blindness or at least slow adaptation. It is often said that eating carrots (or other foods rich in Vitamin A) will improve night vision, but this is true only if you are deficient in Vitamin A to begin with, which is unlikely if you eat a proper diet. The best approach is simply to take care of your eyes and wear sunglasses for protection against glare on bright days. This is important not only for general eye health, but also for the best possible night vision.

One means of preserving night vision after the eyes have become dark-adapted is to close one eye when you are exposed to a bright light. Doing so will greatly reduce the time needed for readaptation. The eye adapts much faster after exposure to red rather than white light. All boats that sail at night should be equipped with a few red-bulbed flashlights and/or fixed lights, to illuminate the chart table and instruments. If red bulbs are unobtainable, you can use red filtered lenses or a white bulb dip such as that made by Sea/Line. You can even paint the bulbs with red fingernail polish.

Needless to say, any crewmember afflicted with night-blindness, color-blindness, or other visual problems should be sure to inform the skipper before standing watch.

## LIGHTS, BINOCULARS, AND OTHER NIGHT EQUIPMENT

A prerequisite for night sailing is a proper set of navigation lights that meet Coast Guard requirements. These are shown in the accompanying illustration. The points (pt.) referred to in the captions are arcs of visibility, one point being 11¼ degrees. The white light forward with a 20-point (225-degree) arc of visibility is called a "masthead light" under the Navigation Rules, but it need not be at the masthead; it can occupy a lower position on the mast, as illustrated. On an auxiliary sailboat, this light is turned on while proceeding under power, but turned off while proceeding under sail. A red sidelight, which is visible from

## Prerequisites to Sailing at Night

UNDER POWER

UNDER SAIL

- 12-pt. white sternlight
- separate red and green sidelights, 10 pts. each
- 20-pt. white light forward

- 12-pt. white sternlight
- separate red and green sidelights, 10 pts. each
- white light forward OFF

OR

OR

- 12-pt. white sternlight
- combination red and green bowlight, 20 pts.
- 20-pt. white light forward

- 12-pt. white sternlight
- combination red and green bowlight, 20 pts.
- white light forward OFF

*Lights for powerboats and sailboats less than 65.6 feet.*

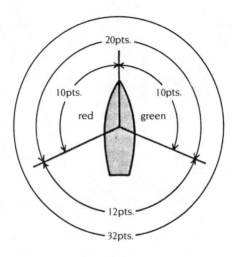

*Arcs of visibility for navigation lights.*

dead ahead to 2 points (22.5 degrees) abaft the beam is displayed on the port side, while a green sidelight with the same arc of visibility is displayed to starboard. The sternlight's arc of visibility faces aft and covers 12 points (135 degrees), 67.5 degrees from dead aft on each side of the vessel.* A power-driven vessel less than 12 meters (39.4 feet) in length may, in lieu of the 20-point white light forward and 12-point sternlight, carry a 32-point (360-degree) white light in addition to her sidelights. If the vessel is less than 23 feet and cannot exceed a speed of 7 knots under power, she need not carry sidelights but must have the 32-point white light. A sailboat less than 23 feet with no auxiliary power needs only an electric torch or

---

*Optional additional mast lights, a 360-degree red above a 360-degree green, may be exhibited in a vertical line near the top of a sailboat's mast. Another option for a sailboat less than 65.6 feet long is a 360-degree light at the masthead which combines the sidelights and sternlight in one lantern. This is carried instead of (not in addition to) the normal lower lights and makes sense for an offshore vessel because the masthead lights cannot be obscured by waves or sails, and they are more visible from the bridge of a ship. Furthermore, one bulb can be used for all three lights, thus saving electricity. I use a Mars–1 unit made by Asimow Engineering that combines a tricolor navigation light, an anchor light, and a strobe flasher.*

lighted lantern showing a white light, but it is safer for her to exhibit the proper running lights as shown in the illustration.

Lights on boats less than 65.6 feet long must be visible for two miles, except that a boat under 39.4 feet needs only one-mile visibility for her sidelights, and a boat 39.4 feet or more needs a three-mile visibility range for her mast light. Further details of navigation lights will be given in Chapter 4.

In addition to proper running lights, there are certain other fixed lights that are virtually indispensable for night sailing. Foremost among these is the compass or binnacle light, which, as mentioned, must have a red bulb to save the helmsman from night-blindness. Instrument dials also need illumination—again with red bulbs. I like to keep most instruments, particularly those that are only occasionally consulted, down low so that they are out of the helmsman's path of vision; those that need a high location should have fairly dim bulbs, or better yet, lights that allow adjustable dimming. Spreader lights are exceedingly useful at times but should not be overused, as they can cause temporary night-blindness. Below decks, the most vital fixed lights are those in the navigation and galley areas. Ideally, the chart table or navigation area should have both a white light and a red one with a swivel neck. Cabin lights must be shielded so that they can't be seen directly by the helmsman.

Equipment needed for night sailing will obviously include a number of flashlights. These should be reasonably distributed throughout the boat. On a boat of moderate size, I would have one in the forward cabin where sails are often stowed, two in the chart table area (one of these being stowed near the companionway so that it can be reached from on deck or below), and two on deck. One of the two deck flashlights should be powerful, perhaps a Halogen light to be used for emergencies or to spot distant objects; the other should be a fairly dim light that can be used to look for deck gear, read a chart, or check sail trim without blinding the helmsman. The latter light needs a red bulb, and I would have at least one of the belowdeck flashlights fitted with a red bulb as well. The flashlights should all be waterproof, or at least water resistant, and they should have lanyard rings or other means by which they can be secured. Avoid cylindrical types that can easily roll when the boat is heeled. I prefer that at least one flashlight have a broad-beam/fine-focus (flood/spot) adjustment switch. If stowage on deck presents a problem, flashlights can often be cached in a winch base locker. The helmsman should be careful that a flashlight stowed near the binnacle will not affect the compass. Never use flashlights with magnetic holders.

Another important tool for night vision is the binocular. In addition to magnifying objects you are looking for, binoculars also gather light, increasing brightness and visibility in the dark. They are commonly designated by two numbers separated by an X (stamped on one of the barrels); for example, 8 X 40. The first number shows the magnifying power; the second provides the diameter of the objective lens (the front lens, furthest from the eye) in millimeters. High magnification and large-diameter objective lenses (for effective light gathering) provide the best ability to see in the dark, but magnification power beyond eight is unsuitable for marine use because the boat's motion makes it too difficult to hold the glasses steady enough to keep an image within the field of view. 8 X 56 binoculars, sometimes called "star gazers," are excellent for calm conditions, but the best all-around performance is provided by 7 X 50s. These are the "night glasses" that have become the standard binoculars for about all the navies of the world.

There are several measures of nighttime effectiveness. The exit pupils, seen as small disks of light on the oculars (near lenses) when the glasses are held at arm's length and aimed at a light source, are a common measure of light transmission. To figure the exit pupil, divide the objective lens by the magnification. A 7 X 50 binocular would have an exit pupil of 50/7 or 7.1 mm. This size correlates well with optimal viewing, since the entrance pupil of the eye can open up to over seven millimeters in total darkness. (The size of the pupil opening, however, decreases with a person's age.)

Other indicators of binocular night performance are the so-called twilight factor and relative light efficiency (RLE). The former is a way of measuring the fact, mentioned earlier, that magnification and objective lens diameter improve efficiency in poor light conditions. The twilight factor is designated as the square root of the power times objective lens diameter; a 7 X 50 binocular would have a factor of $\sqrt{7 \times 50}$ or 18.71. RLE is derived from relative brightness, which is the square of the exit pupil. To determine the relative light efficiency of your binoculars, square the exit pupil and add one-half of that figure—or in other words multiply the relative brightness by 1.5. The RB of a 7 X 50 binocular, for example, is 7.1 squared, or 50.4, and the RLE is 1.5 times 50.4 or 75.6. The RLE figure roughly accounts for the extra effectiveness of magnesium fluoride–coated lenses. Night glasses should have coated lenses for increased light transmission, with an RLE of 54 at the very least and a twilight factor greater than 15.

It is also desirable that night glasses have a reasonably wide field of view, as this facilitates finding objects and minimizes the need to sweep the glasses back and forth over the area being searched. Field of view is the size of the area visible through a binocular and is designated by either a linear or angular measurement. The former measures feet-across-the-scene, viewed from a distance of 1,000 yards; the latter measures the angle subtended at the binocular by the field of view. You can convert feet to degrees by dividing by 52.5. For example, a field of view of 400 feet (about the minimum I would want) converts to an angle of 7.6 degrees.

Essential equipment should of course include a proper

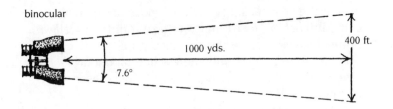

*Binocular field of view.*

compass (mounted in a binnacle if at all possible), a lighted hand-bearing compass, and suitable charts of the waters you sail on. These items as well as other navigation gear will be discussed in Chapter 4. Equipment specifically related to man overboard situations and retrieval at sea will be discussed in Chapter 5.

Every boat, whether or not she sails at night, needs a suitable selection of anchors. For night work I would be sure that the anchor line is marked with high-visibility tape markers (commonly bright yellow with dark red numbers) that can be felt with the hands on very dark nights. A few sailors mark their anchor rodes in a manner similar to a leadline with strips of leather and cloth rags; whatever system you use, the point is to be able to feel the scope in the dark as the rode is payed out. On a small boat in shallow water it is often handy to have a lightweight sounding pole; this should also have marks that can be distinguished by feel. One mark might be of line, another of duct tape, another of reflective tape, and so forth. Your sense of touch can be put to good use in the dark.

2

# Your
# First
# Moonlight
# Sail

To consolidate some of the material in the previous chapter—as well as give you a feeling for night sailing if you've never been out before—let's pretend I'm taking you for your first moonlight sail aboard my 37-foot sloop, *Kelpie*. Welcome aboard! This will be a short night sail, just for the pleasure of it; I will have more to say in Chapter 6 about overnight passages.

Though we'll be leaving from the dock, I did most of the preparation while *Kelpie* was still on her mooring. It might seem odd that I don't like to prepare the boat for sailing under the bright dock lights at the marina, but strong light lengthens the time it takes for our eyes to adapt to the darkness. This may not be of great importance on a bright moonlit night like tonight, but certainly on a dark night you don't want to leave until your eyes are completely adapted. This is especially true when you have to pass through a harbor full of anchored boats and unoccupied moorings, or crab or lobster pots, or channels marked with unlighted buoys. I prefer to prepare the boat in a dark area and use my spreader lights to see what I'm doing. I can turn them off 15 minutes to a half hour before leaving and depart with the best possible night vision. Small boats might use a battery-powered floodlight.

## CHECKLIST

Every boat should have a basic procedural checklist for getting underway, covering such vital items as the condition of the bilge, the position of seacocks, and the amount of engine fuel. At night, of course, a very important check is the condition of our lights. We want to be sure that all the running lights are working, as well as the binnacle light and chart table light. I make it a point to check the state of my two 12 volt, 95 amp batteries, using a permanently installed battery condition meter. Let's be sure to check the flashlights too. I keep some spare C and

D cell batteries for small flashlights that are dim or flickering.

The dodger should be struck for visibility, and the forward hatch closed for safety as well as visibility. We'll remove the mainsail cover, adjust the outhaul, and rig the main halyard so that it is all ready for hoisting.

In rigging the boat for night sailing, we want to doublecheck that all lines are properly coiled and led correctly, and that winch handles are stowed where they belong. On *Kelpie* the jib sheets lead outside the lifelines, and we want to be sure the jib halyard is not fouled around the spinnaker halyard or lift. I've elected to use the number three genoa because there's a moderately fresh breeze, the sail has modest overlap to allow easy tacking, and its luff length is sufficiently short to allow a tack pendant for good visibility. I think we'd better use the extralong tack pendant so that we can see under the sail even when the

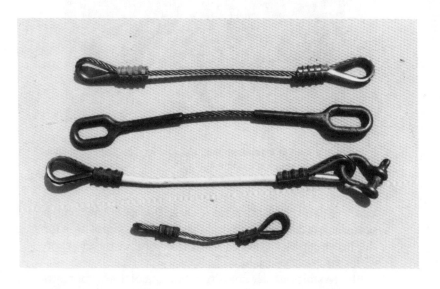

A *selection of tack pendants, some with swaged and others with Nicopressed eyes.*

*A moderately long tack pendant gives the best visibility without unduly compromising sail power at night.*

boat is well heeled when close-hauled. The ship's notebook is consulted to find the proper lead position for the jib sheet blocks when using the number three genoa with extralong pendant.

Before starting the engine to motor out of the harbor into open water, I want to pump the bilge. With the engine off I can hear the pump sucking air when the bilge water drops below the intake hose. All valves or seacocks should be properly turned on or off. A bilge pump's valve should be closed after pumping to prevent back-siphoning while underway. Any sink that is far from the boat's centerline should be closed to prevent inflow, and of course, all scupper seacocks should be open. See that everything is well stowed below, so that it can't fall when the boat is heeled.


## UNDERWAY

We'll allow about 15 minutes in the darkness for eye adaptation, more than enough time to tell the crew what's expected of them. The instructions are simple. Most of the crew should station themselves in the cockpit and keep low so as not to obstruct the helmsman's vision. One trustworthy crewmember takes a flashlight far forward to stand lookout. We'll be weaving our way through mooring buoys on the way out, so the bow watch looks for them and shines the light on any that the helmsman is headed for or passing by too closely. The light is used no more than necessary, and never shined aft into the helmsman's eyes. When you go below to get the flashlights, pass up the air horn so I can stow it near the helm. We might need it to make our presence known to other boats underway.

The engine is started just prior to getting underway, and I'll let it run a few minutes to warm up and to make sure that the oil pressure and exhaust systems are functioning normally.* We also need to check that no lines are hanging

---

*Boats without engines, of course, will have to sail away from the dock and through the harbor before reaching open water. This may sound frightening, but if the boat is a smart sailer and the wind and current are favorable, there is little to fear. In one respect, an engineless boat offers an advantage in allowing her crew superior hearing ability. In a fresh breeze sail must be reduced for conservative speed with adequate steerageway through the harbor and to prevent excessive heeling that might hamper steering and restrict visibility.

overboard that might foul the prop when I put the engine in gear. Superfluous docking lines should be taken in. At night or when shorthanded it is often a good idea to lead the bow and stern lines around the dock's piling and back to the boat so that we can cast off from on board. We are on the leeward side of the dock, so getting underway is simply a matter of casting off. If we were on the windward side, I might want to back down on a springline to push the bow clear of the dock. There should be fenders rigged at strategic locations where the rail might bear against the dock, and a couple of crewmembers should fend off.

Now we're underway. The crew have carefully moved to their stations in the cockpit, and the bow watch is forward. I am standing up at the helm, and we're working our way slowly through the harbor full of boats and moorings. Most of the boats are white and easy to see in the moonlight. The dark hulls are less visible, but they usually have white cabin trunks that stand out well. I give the boats a wide berth because their moorings are often far ahead of their bows. We should also be on the lookout for any dinghies tied astern. The bow watch is shining a flashlight occasionally on a buoy that might be hard for me to see.

Once clear of the harbor, we must negotiate a narrow channel. It is marked with a couple of prominent, well-lighted channel markers. Farther out, though, there are some unlighted buoys for which we have to keep a careful lookout. A few years ago a friend of mine, steering his boat into the harbor on a dark night, hit two of the buoys dead on. It is certainly advantageous to pinpoint the location of the buoys on the chart, so you'll know what to expect and where. Also, it helps to steer a slightly yawing course to be sure the buoys are not dead ahead, hiding behind the mast or cabin trunk. I know from memory what compass course to steer, but in unfamiliar waters I would want to lay out a safe course on the chart. A particular hazard in our region of the Chesapeake is crab pots; if possible, we should try to avoid those areas where they are allowed, which are marked by unlighted white buoys with orange markings.

## IN OPEN WATER

At last we're out in the open water and can relax a bit—
though of course, we still have to keep a careful watch for
other boats. The helmsman is the one primarily responsible
for avoiding collisions, but there should also be at least
one other lookout to watch certain spots that are difficult
for the helmsman to see, such as the one behind the jib.
Keep a sharp lookout for small rowboats and sailboats
(under 23 feet), which need show only a flashlight or
lantern.

What say we make sail? Let's take the stops off the main,
slow the engine, and head into the wind. An experienced
crewmember can work the main halyard winch, using the
tape marker on the halyard as a guide to how far to hoist. I'll
slack the mainsheet and see that it's clear to run. Now we
can bear off and cut the engine. The jib goes up next. On
small boats the jib can usually be hoisted from the cockpit,
but on *Kelpie* the halyard cleat is on the mast. Careful up
forward! Move cautiously, and be sure to lock on the winch
handle. The cockpit crew can trim in the jib after it is fully
hoisted. Isn't it peaceful without the engine?

**Getting the feel upwind**—Care to take the helm? The
breeze is just right for a first-time moonlight sail—not too
light nor too fresh. Never mind right now about steering by
the compass; just pick out a star or a light on the horizon to
head for. You could even steer for that dark clump of trees
on the shoreline. Keep your head up so that you can better
feel the wind on your face and neck. That gives you a good
indication of its direction and strength.

Let's put *Kelpie* on the wind. Head up slowly toward the
wind while we harden in on the sheets—but not too high,
since that would kill her way. Now that we have the sails
trimmed about right, we can sail her almost entirely by feel.
You can sense when she's hit by a puff. The wind feels
cooler on your face, the angle of heel increases, and there
is more weather helm. Let her come up a little until the
helm slackens. That's too much; you can hear the sails

21

luffing, and it feels as if we're slowing down. Bear off a little—that's it—now you have her "in the groove." You can sense her speed partly by the sound of the water. It makes a steady sloush, sloush as we power into the seas. The sound diminishes if we slow down.

Now that you've gotten the feel of it, you can glance at the instruments. The knotmeter shows we're making five knots. Seems as if we're going a lot faster, doesn't it? When you suspect we're slowing down and need to bear off a bit to get more speed, glance at the knotmeter again to verify. You should also check the compass occasionally to be sure you're not too far off the wind and to see if you're being headed or lifted. Easy on the helm—you don't want to oversteer.

**The sky**—Look at the moon and all those stars. The moon is almost full. Is it waxing or waning? One way to tell is to see which side the shadow is on. A handy remembrance aid is to think that when the moon crescent makes a C with the shadow on the right, it is ceasing or waning.

Notice how many more stars you can see in the open water where there are no trees, buildings, or glow of lights to obstruct your view of the sky. In trying to identify stars, I usually look first to the north and find the constellation Ursa Major, the Big Dipper. The two end stars of the dipper's bowl, the "pointers," are aimed at Polaris, also known as the North Star. The distance to Polaris from the end pointer is about four and a half times the distance between the pointers. Other constellations near Polaris are Ursa Minor (the Little Dipper), Draco (the serpentine constellation), Cepheus (a five-star cluster looking like a steeple), and Cassiopeia (resembling an M or a W). These constellations are circumpolar for us at our latitude (about 40 degrees north)—that is, they never rise or set but simply seem to rotate as the earth turns on its axis.

**Off the wind**—Well, back to sailing. Now that we've had some practice sailing to windward, let's try running off.

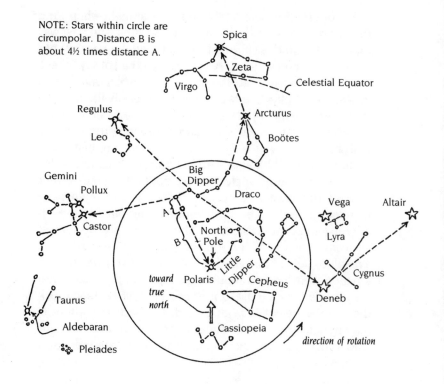

NOTE: Stars within circle are circumpolar. Distance B is about 4½ times distance A.

Spica
Zeta
Virgo
Celestial Equator
Regulus
Arcturus
Leo
Boötes
Gemini
Big Dipper
Pollux
Draco
Vega    Altair
A
Castor
North Pole
Lyra
B
Little Dipper
toward true north    Polaris    Cepheus
Taurus
Cygnus
Deneb
Aldebaran
Cassiopeia
Pleiades
direction of rotation

*A rough guide to locating the stars.*

Bear off slowly while we ease the sheets. On this point of sail the challenge is to avoid sailing by the lee, which can result in an accidental jibe. Sailing by the lee occurs when you allow the wind to get more than 180° astern, so that it is actually coming over what should be your lee quarter—a precarious situation that can send the boom slamming over with sufficient force to injure any crewmember in its path, or even to capsize a small boat. If we were going to run almost dead before the wind, I'd rig a jibing preventer by leading the boom vang to a point forward of the boom, then snapping the vang tackle's shackle to the base of a lifeline stanchion or some other sturdy fitting.

Accidental jibes are a particular risk at night when wind direction is harder to judge from visual clues. The best way to tell if the wind is shifting is to feel it on the back of your neck, or even on your hair. Of course, you can always check the wind indicator at the masthead, which has been marked with reflective tape so as to be easily illuminated from our masthead navigation lights. Many boats have apparent wind indicators that show the wind direction on a dial at deck level. *Kelpie* has very little sophisticated equipment, but I like her simple Windex indicator with its two reference arms angled aft at about 30 degrees either side of the boat's centerline. To keep the main from blanketing the jib, keep the Windex vane outside the 60-degree gap between the arms. This will also ensure against an accidental jibe. When you see the jib start to go limp as a result of being blanketed, it is a sign that you are straying too far downwind and should head up a bit. If you can't see the jib, you can hear the rubber-coated jibsheet lead block drop to the deck with a dull thud. This indicates there is not enough strain on the sheet to hold it up. Even without lead blocks, other sounds can warn you that the jib is collapsing.

## RETURNING TO PORT

Personally, I could stay out here all night, but I know we all have to get back, and it will take almost as long to return as it did to get out here. So perhaps we'd best come about and head for home. Let's be sure that everything is ready before tacking. Wrap the windward sheet a few times around its winch, move the handle up to windward, and uncleat the leeward sheet. Be sure the jibsheet is free to run, so that it doesn't foul on the mast when we tack. Ready about . . . hard alee! Ease the boat into the wind, then put the helm hard over to make the jib blow around the mast in a hurry. Don't bear off too far on the new tack. It's a good idea to check the compass so that you don't bear off much below a 90-degree angle to the original tack. At night, with fewer visual reference points, your sense of direction can play tricks on you. When the boat swings through the eye of the wind, haul in on the sheet hand over hand as fast as you can before the sail fills; this saves a lot of winching later. Actually, you don't have to get the jib in very flat, as our course for home is a close reach. You can head for that white flasher dead ahead. Shine the flashlight on the jib telltales to be sure the sail is trimmed properly. The leeward yarn is stalled—twirling around rather than trailing aft—so let's ease the sheet a bit.

Now that we're in the river headed for the harbor, we can start the engine and lower sail—first making sure, of course, that there are no lines overboard that might foul the prop. OK, the engine is running. Let's luff her into the wind and lower the jib. Heading up into the wind centers the jib over the foredeck so that it won't drop overboard or force a crewmember to lean outboard of the lifelines to pull the sail in as it comes down. At night the crew must be careful to keep well inboard—for obvious reasons. If we were running for home with a fair wind, another option would be to bear off far enough to blanket the jib. Then we could pull it in over the center of the foredeck once the sheet has been eased.

It's even more important to lower the main while reasonably close to the wind, because otherwise it can foul on the shrouds and spreaders. I've broken battens trying to pull the sail down while it was blown against the rigging—and of course, all such problems are further complicated by darkness. Live and learn.

Powering back through the harbor doesn't seem so difficult as it did coming out. Maybe we have become more accustomed to the darkness, or perhaps the greater altitude of the moon makes it brighter. Although we aren't able to use the reflected path of the moon to light our way through the harbor, we can follow the reflected path of that bright dock light, which shows up unoccupied mooring buoys. We'll land at the well-lighted part of the dock

25

because eye adaptation is no longer a great problem after landing. Guess I'll come in on the windward side, since the dock is well-fendered and there's not too much breeze. A windward landing is easier at night, since misjudgments of distance or unseen wind puffs will work in your favor, pushing you down on the dock rather than farther away from it. Get the docking lines out, and stand by to fend off. Here we are back again, and we didn't get into any trouble at all, did we?

Hope you enjoyed our imaginary sail. I've just touched on many of the details of sail handling, helmsmanship, and safety. In the following chapters we will cover these matters much more thoroughly.

# 3

# Helmsmanship, Deck Work, and Sail Handling

## HELMSMANSHIP

Chapter 2 touched on the art of sailing by feel, but a little more should be said about making full use of the senses to steer upwind at night. During the day, the helmsman sails to windward primarily by using his sense of sight. He watches the luffs of sails and particularly the luff telltales. After the sun goes down, other senses must be more fully utilized. "Feel," or what is sometimes called the "tiller touch," is really a composite of awarenesses: the sense of balance (awareness of heeling angle), the feel of pressure on the helm (weather, lee, or slack helm), the sensation of wind on your skin, the sounds of the water and sails flapping, the sense of motion when the boat pitches in head seas and accelerates or decelerates from the effects of puffs or waves.

To sail upwind effectively at night, you need to develop a feel for when your boat is sailing "in the groove"—at peak efficiency. It is relatively easy to tell when you are on the

27

"In the groove" as darkness approaches.

high (upwind) side of the groove because the boat noticeably slows down, she heels less, the sound of her bow wave diminishes, the shaking luffs are audible, and on some boats there is even a noticeable vibration of the rigging or mast. The low side of the groove (sailing too far off the wind) is much more difficult to detect; there is no flapping of sail luffs or diminishing of bow wave to hear, and the boat feels as though she is sailing very fast. (She is indeed making knots, but she is not making optimal progress to windward.) There are several important clues that indicate you are sailing too low: the boat tends to heel more than she should, she may have additional weather helm, and the wind feels as though it is blowing more on the side of your face. Some experienced night sailors can even feel the hairs on the tops of their hands shift direction and thereby act as telltales. Other sailors can detect wind shifts by hearing differences in sound when the wind changes its angle against their ears. To doublecheck that you are not sailing too low with stalled sails, it's a good idea to test the wind periodically by trying to head up. When you feel the boat straighten up, slow down, or become slack on her helm, ease off again until you feel her heel a bit more and pick up speed. Bear in mind that winds at night, being cooler, are generally heavier than winds during the day, and this extra weight may slightly increase the angle of heel.

It might be appropriate to mention here the not uncommon "night wind." This is an offshore breeze that may occur several hours after sunset when the land becomes considerably cooler than the water. These winds can be strengthened by sloping coastal hills that cause sinking air to flow seaward. When there has been a sea breeze during the late afternoon, a temporary calm may be produced after the land cools off and just before the beginning of the night wind. Under these circumstances the night sailor will often profit by hugging the shore to pick up the initial and strongest component of the land breeze.

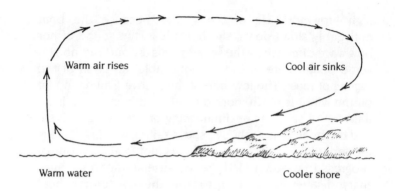

Warm air rises

Cool air sinks

Warm water

Cooler shore

*Basic principle of the night wind.*

Instruments are also valuable in helping the helmsman stay in the groove. Basic instrumentation for sailing to windward should include a knotmeter (speedometer), an inclinometer (to show angle of heel), and an apparent wind indicator. The latter may be the electronic kind that shows the wind angle on a dial at deck level, or it can be a simple illuminated masthead wind vane that is read by looking aloft. Among the various brands of the latter, I find the Windex vane particularly helpful in avoiding the low side of the groove, since it has two projecting arms (one for each tack) that show the approximate ideal close-hauled sailing angle. Instruments should not be continually stared at, but only monitored periodically to check, reinforce, or augment sensory perceptions.

The compass is often vital to helmsmanship. On an upwind course at night one of its most useful functions is to help you detect wind shifts you might not spot by visual clues. If you find yourself sailing into a severe header (adverse shift), it might very well pay to come about and take advantage of the lift on the opposite tack. Most of the time, though, particularly on long passages offshore, the compass is used primarily for steering compass headings

*The Windex vane is particularly helpful for upwind sailing at night, since its two projecting arms show the optimum close-hauled sailing angle. Some models are available with masthead lights. (Courtesy Davis Instruments Corporation)*

on reaching courses. It is important that the navigator keep a record of the headings for a reasonably accurate dead reckoning position; this topic as well as more detailed information on compass installation will be covered in Chapter 4.

When steering by compass at night, particularly in the middle of a long trick at the helm, it is easy to become disoriented and to think of the boat as being steady and the compass as moving, when actually it is the other way around. The inexperienced (or weary) helmsman should constantly remind himself that the compass is fixed, and the lubber line must be aligned with the compass degree mark or point representing the proper course. Another problem that can easily become exaggerated at night is a tendency to oversteer. When the boat starts to yaw, she should be eased back on course, and the helm should be gently reversed just before the lubber line is aligned, in order to "catch" the swing before the boat yaws in the opposite direction. Practice may not make perfect, but it will definitely improve.

## DECK WORK AND SAIL HANDLING

Advise your crewmembers to become "nightcrawlers" after sunset. Although they might get away with rushing about the boat in the daylight, they should move with careful deliberation in the dark. It is usually advisable to keep low with your feet wide apart for good balance and to kneel when handling sails on the bow. When the boat is heeled, move forward or aft along the windward side deck, and in rough weather particularly, remember the adage "one hand for the ship and one for yourself."

It is important as well to exercise utmost care in handling lines and winches in the dark. See that every line is properly coiled and securely hung when not in use. Periodically, untwist your lines to chase out kinks. This may be done by turning or spinning the line in your hand while coiling, or in stubborn cases, by trailing the line overboard for a short while (be sure the engine is not running so the line can't tangle around the prop). Some modern lines that are braided or plaited rather than laid (of twisted construction) are less prone to kinking when coiled in a figure eight fashion. Never stand in the coil or the bight of a line, and learn to tie the basic knots (bowline, square knot, and half hitches) in the dark.

When winching, watch out for overriding turns of the line on the winch drum, and keep your head well clear of any winch handle. Also, keep your fingers away from the turns around the winch, and use the heel of your hand to apply some pressure to the turns when slacking off so that the line will not slip too suddenly. Be sure that wire-servicing reel winches with screw toggle brakes are operated with the brake on, and never slack off with the handle inserted. Hoist-limiting marks on halyards should be of tape or twine that can be felt in the dark. No crewmember should be allowed to handle running rigging and its gear in the dark until familiar with its location and operation in broad daylight.

Sail trim can be tedious in the dark, although less so on a cruiser than a racer, which must have near perfect trim every moment. It pays to use a dim flashlight periodically to check that the luffs are kept on the verge of luffing or bulging inward. Some racing boats carry trim lights, small spotlights mounted on the mast or rigging to shine on luff telltales, but some of these can cause night-blindness to an unacceptable degree. Luff telltales should be kept streaming aft (not twirling around) on each side of a sail to assure that it is neither luffing nor stalled. Most people find that black telltales on white sails are easiest to see at night. Some boats even have black bands or circular patches marking reef points and grommets for greatest contrast in the dark. I find it helpful to have a black tape or paint reference mark on the spreader at a distance inboard of its tip equal to the distance outboard from the tip to the genoa leech when the jib is correctly trimmed for beating to windward. Then the leech's proper distance off the tip can be more easily judged in the dark by comparison. Correct positions for sheet leads should be marked as boldly as possible. Adjustable rigging such as backstays should also be marked with tape at optimum points of adjustment. When you use a tack pendant at night to improve visibility, remember that this will move the sheet lead farther aft, and its position should be plainly marked.

Headsail changes at night are most easily accomplished

when the boat is running off. With the wind aft there is less heeling and violent motion, and the forward sails are blanketed, enabling easy handling. On a racer, obviously, it is not always possible to change course, but on a cruiser the helmsman is usually able to bear off and head downwind for a few moments while the headsail is changed. When the boat is equipped with hanked-on jibs, an effective plan is to carry a small jib furled and stopped to the rail but hanked to the forestay beneath the larger jib that is being flown. The small jib can be rigged in this manner while it is still daylight, and if headsail reduction is necessary after dark, all you need do is lower the large jib, unsnapping as it comes down, transfer the halyard, and hoist the smaller replacement. With luff foils, which support jibs by their luff bolt ropes in full-length grooves, a headsail change is much more difficult because luffs are partly or completely detached during the lowering and hoisting operations. On a shorthanded offshore boat I would not have such a system unless it is a roller-furling luff foil. In my opinion, cutters or other double head–rigged boats should always have their forestaysails supported by wire stays that can accept hanks.

For mainsail reefing in the dark I prefer the jiffy system, whereby the halyard is slacked and a luff grommet above the tack is either hauled down with a reefing line or hooked onto a tack hook. Next, the halyard is tightened, and the clew is winched close to the boom with a leech earing (reefing line). A simple method of rigging the leech earing is to run it from an eye strap on the boom under a leech cringle (grommet) up through the grommet and down to a cheek block near the end of the boom. Then run it forward along the boom through fairleads and blocks to the base of the mast and either to a mast winch or back to a cockpit winch. Put a twine or tape mark on the halyard, indicating the amount of halyard that must be slacked off to lower the tack to its reefed position. You might want to place a similar mark on the leech earing to show when the line is hauled sufficiently taut to bring the clew cringle down to

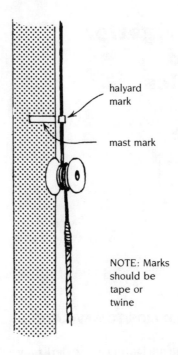

halyard
mark

mast mark

NOTE: Marks
should be
tape or
twine

*Marking the main halyard so that it can be felt in the dark.*

the boom. When shortening down at sea with a small crew, the spreader lights can be useful, but when there is a full crew and the boat is in crowded waters, I like to send a crewmember forward with a red-bulbed flashlight in order to best preserve the helmsman's night vision.

# 4

# Navigation and Rules of the Road

## LIGHTED AIDS TO NAVIGATION

Although navigating in the dark may at first seem intimidating, in one important respect piloting is actually easier at night: lighted aids to navigation are more easily seen from a distance after the sun goes down, and also the loom of shore lights or lighthouses may be seen on low-lying clouds. This is why experienced passagemakers often prefer to make landfalls at night, or just before dawn, so as to enter the harbor itself by the breaking light of day.

Lighted buoys, channel markers, and other aids to navigation have distinctive light signals that are readily identified by consulting charts of your area or the *Light Lists* published annually by the U.S. Coast Guard (available from the Superintendent of Documents, U.S. Government Printing Office, Washington, D.C. 20402, or at many local chart outlets). The *Light Lists* show the characteristic light rhythms of the various signals and explain the terminology and abbreviated symbols used on the charts—as, for example, the difference between quick flashing, group

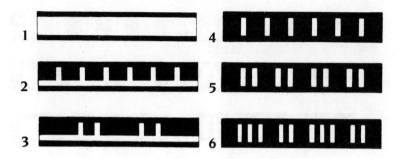

| | SYMBOLS AND MEANINGS | | |
|---|---|---|---|
| | **Lights which do not change color** | **Lights which show color variations** | **PHASE DESCRIPTION** |
| **1** | **F.**=Fixed ------------ | **Alt.**=Alternating. | A continuous steady light. |
| **2** | **F.Fl.**=Fixed and flashing. | **Alt.F.Fl.**=Alternating fixed and flashing. | A fixed light varied at regular intervals by a single flash of greater brilliance. |
| **3** | **F.Gp.Fl.**=Fixed and group flashing. | **Alt.F.Gp.Fl.**=Alternating fixed and group flashing. | A fixed light varied at regular intervals by groups of 2 or more flashes of greater brilliance. The group may, or may not, be preceded and followed by an eclipse. |
| **4** | **Fl.**=Flashing-------- | **Alt.Fl.**=Alternating flashing. | Showing a single flash at regular intervals, the duration of light always being less than the duration of darkness. |
| **5** | **Gp.Fl.**=Group flashing. | **Alt.Gp.Fl.**=Alternating group flashing. | Showing at regular intervals groups of 2 or more flashes. |
| **6** | **Gp.Fl.(3+2)**=Composite group flashing. | ------------------------ | Group flashing in which the flashes are combined in alternate groups of different numbers. |

*Characteristic light rhythms for marine navigation aids. (From the Coast Guard Light Lists)*

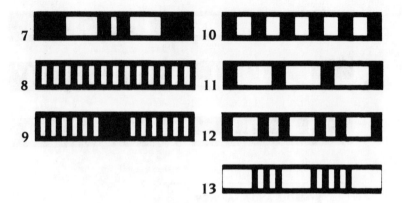

| SYMBOLS AND MEANINGS | | |
|---|---|---|
| **Lights which do not change color** | **Lights which show color variations** | **PHASE DESCRIPTION** |
| 7 **Mo.(K)**=Morse Code. | ---------------- | Light in which flashes of different durations are grouped to produce a Morse *character* or *characters*. |
| 8 **Qk.Fl.**=Quick flashing. | ---------------- | Shows not less than 60 flashes per minute. |
| 9 **Int.(I)Qk.Fl.**= Interrupted quick flashing. | ---------------- | Shows quick flashes for about 4 seconds, followed by a dark period of about 4 seconds. |
| 10 **Iso.(E.Int.)**=Equal intervals. | ---------------- | Duration of light equal to that of darkness. |
| 11 **Occ.**=Occulting. | **Alt.Occ.**=Alternating occulting. | A light totally eclipsed at regular intervals, the duration of light always greater than the duration of darkness. |
| 12 **Gp.Occ.**=Group occulting. | **Alt.Gp.Occ.**=Alternating group occulting. | A light with a group of 2 or more eclipses at regular intervals. |
| 13 **Gp.Occ.(3+4)**= Composite group occulting. | ---------------- | Group occulting in which the occultations are combined in alternate groups of different numbers. |

flashing, occulting, or Morse code flashing. Aids to navigation are often color-coded as well (normally green or white for black markers and red or white for white markers), and this information is also designated in abbreviated form on your chart and in the *Light Lists*.

A stopwatch is a useful tool in timing flashes, although you can always say "one chimpanzee, two chimpanzee..." to approximate seconds. In unprotected waters, be careful not to become confused by interrupted signals resulting from lights being obscured intermittently by waves.

## CHARTS AND CHART TABLE

Your basic tool for all navigational purposes—daytime or nighttime, local cruise or long passage—is your National Ocean Survey chart. Normally this will be an intermediate-scale (about 1:40,000) chart (or charts) of your cruising waters, but for entering and leaving anchorages you may also want to carry larger-scale (1:20,000 or even larger) harbor charts. Local charts are available at retail outlets in most coastal cities and towns, but for planning a long passage—or being sure that the chart you need is in stock—it's a good idea to purchase your charts by mail directly from the Distribution Branch N/CG33, National Ocean Service, 6501 Lafayette Ave., Riverdale, MD 20737. For most areas of the United States you can also obtain *Chart Kits*: large, spiral-bound books published by the Better Boating Association and containing a comprehensive collection of charts for a given geographical area.

Before leaving an unfamiliar anchorage at night, you should plot (using parallel rules or other plotter) your course on the chart to provide compass headings. Then check off, mentally at least, all the navigational aids as you pass them. This will help ensure that you don't become lost when there are a number of shore lights forming a confusing background behind channel markers or buoys.

Chart work is made easier by a suitable work area. Small

open boats might have a portable drawing board to which a chart can be fastened. On small cruisers the work area can be a galley counter, the top of the icebox, or a dinette table, but on a larger boat a proper chart table is preferable. "Proper" means a table at least 30 by 30 inches, having at least one shallow drawer or a large, flat compartment with a lift-up top to store folded charts. Remember that there should be both a moderately strong white light and a red light to help night vision. Shelf space is needed around the chart table for such tools as radios, Loran, books, and various navigation instruments.

The chart table should be located close to the companionway so the navigator can move quickly from his work area to the deck and communicate easily with the helmsman, but far enough from the hatch to be protected from spray or rain. Sometimes a curtain can be rigged both to ward off spray and to help shield the helmsman from the navigator's light.

## THE COMPASS

Your compass is your most vital instrument for low-visibility navigation. Even small open boats should have one that is suitable both for steering and taking bearings. Don't try to economize with a cheap compass. Be sure it is accurate, has an easily readable card, and is located where the helmsman can always see it. I prefer a single compass on a pedestal-type binnacle directly forward of the helm, but a pair (one on each side of the boat) is fine as long as the two closely agree. Most sailors find that the most readable compass card is one labeled with large degree numbers every 30 degrees and with 10- and 5-degree intervals indicated by longer and shorter radial marks. Again, the binnacle should be fitted with a red light.

A pedestal binnacle facilitates taking bearings on objects, since you can sight directly across the top of the compass. Bearings by this method are less precise but still

*Kelpie's compass, mounted on a pedestal binnacle. The compass card, marked in 5-degree intervals with large degree numbers every 30 degrees, is easy to read at night.*

possible if the compass has a dome-shaped glass (which is preferable because it magnifies the card). For greater accuracy, you should also have a hand-bearing compass that can be held up to your eye; again, the two compasses should closely agree. The hand-bearer also serves as a spare compass, something every cruiser should have, and it can even be used as a "telltale," a special belowdeck compass that is usually mounted upside down over the skipper's or navigator's bunk. My boat carries a Weems and Plath hand-bearer mounted (right side up) in a rack over the chart table, and from the quarter berth, just abaft the chart table, I can see the boat's heading on the instrument's magnifying prism. On long passages it is comforting for the resting navigator to be able to glance at his telltale occasionally to be sure the helmsman is on course.

When installing a compass, be sure to consider the effects of deviation, the error caused by magnetic attraction aboard the vessel. Unacceptable deviation usually results from placing the compass too close to large ferrous objects or electrical equipment. A rule of thumb is that the compass should be at least three feet away from electronic gear and six feet from the engine. Any electrical wires near the compass should be twisted, with each conductor wound around the other to cancel their magnetic fields. Be wary of any metal or electronic object placed near the compass. Deviation has been caused by winch handles, beer cans, steel-rimmed eye glasses, photographic light meters, and even nickel plating on the binnacle. When errors cannot be entirely removed, you need a deviation card to show the errors on all headings. "Swinging the compass" to construct such a card is outside the scope of this book, but is covered in *Sailing in the Fog* by Roger Duncan, a companion volume in this Seamanship Series. It is an inexact science, so for greatest accuracy, have the compass adjusted by a professional adjustor.

Be sure the compass you purchase is intended for a sailboat. Powerboat compasses may not be designed to keep their cards level at high angles of heel. And take care as well that the compass is installed with the lubber line (reference mark or post on the bowl that indicates the boat's heading) exactly parallel to the boat's centerline—otherwise you will have a guaranteed source of inaccuracy.

## PILOTING AND DEAD RECKONING

In its simplest form, piloting is a matter of finding your position by taking bearings on visual objects shown on the chart. At night, obviously, bearings are taken on lighted objects—normally aids to navigation but sometimes other things, such as bridge lights or radio towers. The easiest way to take a bearing is to sight over the top of the steering compass or to use a hand-bearing compass. A compass bearing provides a line of position (LOP), a line drawn on the chart along the bearing you have taken through the object sighted. The boat must be located somewhere on this line, but you don't know exactly where. To obtain a fix— that is, to pinpoint the boat's position—you need two or more LOPs crossing each other at fairly wide angles. This requires a minimum of two bearings, quite far apart. It is really best to have three LOPs, since bearings are almost never completely accurate. A small triangle called a "cocked hat" is formed where the three lines cross, and you can assume that you are located in the middle of the triangle.

When there is only one distant light on which you can take a bearing, and your boat is passing it (not headed toward or away from it), you can nevertheless estimate your probable position using dead reckoning (DR), a method based on keeping track of your course and the distance traveled. There are three common ways of doing this: a running fix, a bow-beam bearing, or a method known as doubling the angle on the bow.

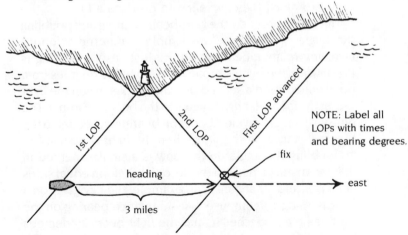

*Running fix using one object.*

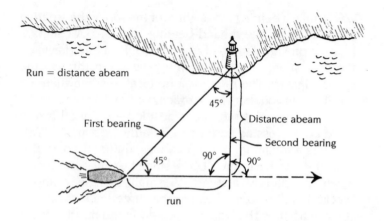

*Bow-beam bearing.*

The running fix involves taking a bearing on a distant light, then taking a second bearing on the same light sometime later and advancing the first LOP along the course to cross the second LOP. For example, a boat moves past a lighthouse in an easterly direction for a distance of three miles between bearings. After the second LOP is drawn on the chart, the first LOP is moved three miles eastward along the courseline to produce a fix.

The other methods, the bow-beam bearing and doubling the angle on the bow, are simple geometric solutions involving relative bearings. In the first case, a bearing is taken when the light is broad on the bow (45 degrees from dead ahead), and a second bearing is taken when the light is exactly abeam (at right angles to the boat's heading). The run, or distance traveled between bearings, is equal to the boat's distance off the light when the light is abeam.

Doubling the angle on the bow is a similar method in that the distance run and the second LOP form equal sides of an isosceles triangle. As an example, the navigator on a boat passing a lighthouse takes a relative bearing on the light as he approaches it, and the light bears 25 degrees from dead ahead. The boat continues on a straight course until another bearing double the angle of the first (50

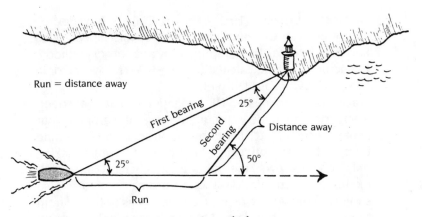

Run = distance away

First bearing

Second bearing

25°

Distance away

25°

50°

Run

*Doubling the angle on the bow.*

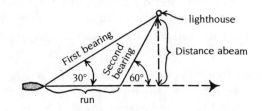

— lighthouse

First bearing

Second bearing

Distance abeam

30°

60°

run

Distance abeam =
⅞ of run when
bearings are
30° and 60°

*Seven-eighths rule.*

degrees) can be taken. At that point two legs of the triangle
are equal, and the run is equal to the distance off the light
at the time of the second bearing. A special form of
doubling the angle is known as the seven-eighths rule. In
this case the first bearing is taken at 30 degrees and the
second at 60. This gives the additional information that
when the light comes abeam, the distance off will be
seven-eighths the distance between bearings.

For any of these methods to be effective, you need to
keep track of your run—the course and distance traveled.
Distance may be determined from a distance-recording
log (either a hull-mounted electronic device or a taffrail log
with rotator towed astern), but if you don't have one, it can
also be calculated according to the simple formula,

45

Distance = Speed × Time (D = ST). Time is provided by your watch—an essential item for all sailors. Speed (also known as rate) can be determined by a variety of methods; the accuracy of your calculation here is a decisive factor in the accuracy of your DR plot.

The great weakness in DR navigation is that boat speed is speed through the water and not over the ground. Your knotmeter may read 5 knots, but if you are at the same time being set by an opposing 1-knot current, you are making only 4 knots over the ground. Mechanical and hull-mounted electronic distance-recording logs are also subject to this source of error, and readings obtained from these instruments will need to be adjusted for leeway and current. The former must be estimated for various conditions after you have become familiar with your boat's sailing performance. Current can be figured from current tables* or simply by watching the flow past buoys and other stationary objects when visibility permits.

One rough method of figuring speed over the ground is to time the passing of a nearby buoy. First, you need the boat's length from bow to helmsman's seat. A person on the bow shouts when the buoy is abeam from his position, at which instant the helmsman starts a stopwatch, stopping it when the buoy is abeam from him. This time is divided into a numerical constant to obtain speed over the ground in nautical miles per hour. The constant for your particular boat is derived by multiplying 0.5925 times the distance from bow to helmsman. For example, if on your boat that distance is 26 feet, your numerical constant is 15.4. If it takes 5 seconds to pass from bow abeam to helm abeam, your speed is 15.4 divided by 5, or 3.1 knots (nautical miles per hour). This method can be particularly useful at night when you are passing close by a lighted aid to navigation.

---

*Tidal current tables are available from the Superintendent of Documents, Government Printing Office, Washington, D.C. 20402. The Eldridge Tide and Pilot Book, published annually by Robert E. White, 64 Commercial Wharf, Boston, MA 02110, has a wealth of information for East Coast mariners.

A quick and easy means of determining distance is by the six minute rule: the distance a boat travels in six minutes equals her speed divided by 10.

The major purpose of piloting, of course, is not only to find out where you are, but to plan where you are going, and to advance your position along an intended course already plotted on your chart. Fixes, supplemented by DR positions obtained by the methods just described, give you a picture of how you are doing and should be carefully noted. It goes without saying that DR positions should be updated by actual fixes whenever possible.

Since the course plotted on your chart is a course over the ground, while the actual course sailed is through the water, one of the important skills required in piloting is to be able to adjust your intended course for current and leeway. When the current is on your beam and you must allow for its sideways push, a handy rule of thumb to allow for the boat's deflection is the 60 formula. Sixty is divided by the length of the course, and the result is multiplied by the distance you are set by the beam current. If you figure the current will set you 3 miles east on a 12-mile course due south, for example, divide 60 by 12 and multiply the resulting 5 by 3; you must steer 15 degrees west of south to make good a course of due south. For a current set of 45 degrees from the course, allow two-thirds of the 60 formula result, and for a set that is 30 degrees from the course, allow one-half of the 60 formula result.

## ELECTRONIC NAVIGATION

There is not room in this book to cover all the electronic instruments and systems used for navigation today. This subject is treated thoroughly in a companion book in the International Marine Seamanship Series, *Piloting with Electronics* by Luke Melton. But I would like to comment briefly on the three electronic navigation instruments most

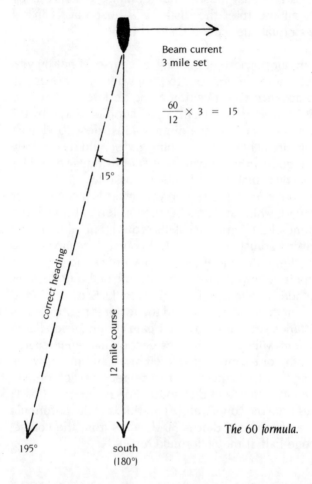

Beam current
3 mile set

$$\frac{60}{12} \times 3 = 15$$

15°

correct heading

12 mile course

*The 60 formula.*

195°

south
(180°)

commonly used, particularly at night, on small to medium-sized American coastal cruisers: the radio direction finder (RDF), the depth sounder, and Loran-C. The first two are relatively inexpensive, and the latter is rapidly becoming so with advances in microprocessor technology.

**Loran-C**—Deriving its name from "long range navigation," Loran is a pulsed, low-frequency radio system that provides hyperbolic (curved) lines of position. The system uses a chain of land-based transmitters consisting of a master station and two or more secondary stations. The master and its secondaries transmit synchronized pulses at precise time intervals, and the shipboard Loran-C receiver measures the time differences (TDs) in microseconds (millionths of a second) between the arrival of each of these pulse groups. A TD between the master and one secondary can be read as a five-digit number display on the Loran receiver and then plotted on a Loran-C chart as an LOP. Another TD from another secondary is read and plotted to obtain a fix.

Loran-C charts are similar to regular charts except that they are covered with a grid of Loran-C LOPs (lines along which the TDs are constant) labeled with four- and five-digit numbers identifying the chain and TDs. For the most accurate fix, you should select lines that cross each other at wide angles—at least 30 degrees or greater. After tuning into the appropriate Group Repetition Interval (GRI, the four-digit number identifying a Loran chain), you obtain TDs of the individual secondaries as displayed on your Loran set. Look on the chart grids for numbers that are close to the readouts. Almost inevitably a displayed number will fall between two of the grid lines, requiring you to draw your LOP somewhere between the two printed lines and parallel to them. Just where you plot the LOP must be determined by interpolation, using dividers and the interpolation scale printed on the chart. After repeating this process with at least one other LOP that crosses the first at a wide angle, you have a fix where your plotted lines cross.

Most modern receivers can give you latitude and longitude displays as well as TDs, which allows you to locate your position on a chart that has not been overprinted with Loran LOPs. Normally the TDs are more accurate, however, and you should carry Loran charts if for no other reason than to confirm your lat/lon position.

Loran-C can be extremely accurate, and unlike dead

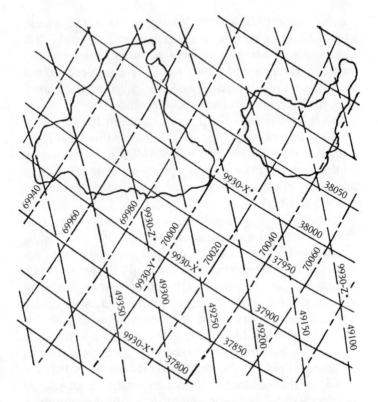

*Sample Loran-C chart showing TDs. (Based on an example in* Loran-C User Handbook, *published by the* U.S. Coast Guard)

reckoning, it is unaffected by current and leeway. No system of navigation should be considered infallible, however. Inaccuracies in Loran may be caused by a faulty receiver, thunderstorms, interference from powerful radio transmitters on shore (on frequencies near 100 kHz), and even slight errors in the Loran TD lines printed on the chart. Furthermore, significant inaccuracies often occur when the Loran signals pass over land. If you look at a Loran chart, you will notice that the TD grids usually stop before running into inland harbors or bays, and it can be a dangerous practice to extend those printed lines yourself. Don't accept your Loran position as gospel; always check it against positions derived from other navigation systems.

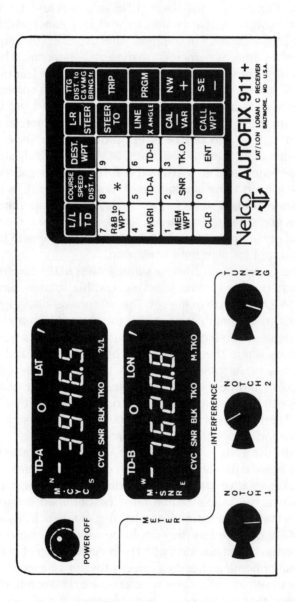

Front panel of an Autofix 911 Loran receiver. Latitude and longitude display showing: north 39° 46.5', west 76° 20.8'. (Courtesy Nelco)

**RDF**—Compared with Loran, RDF is an ancient form of radio navigation that is simpler and less expensive but also less accurate. For the most effective use, you should be in an area of numerous marine radiobeacons. These beacons, whose locations are found in the *Light Lists* or on charts, have characteristic identifying signals, a combination of dots and dashes in Morse code. The *Light Lists* explain the system of transmission. Any commercial AM radio station also can be used, provided you know the location of its transmitting tower. RDF is a relatively short-range system and is most suitable for coastal navigation, as well as for making landfalls after long passages. It is less accurate at night than during the day and is inaccurate at dawn and sunset (times when its use should be avoided), but if these limitations are borne in mind, it can still be a useful tool for nighttime navigation.

The typical small cruising sailboat with RDF will carry a battery-powered portable receiver with a rotating ferrite bar antenna on top of the set. The antenna reads against a 360-degree azimuth scale to give the direction from which the signal is coming. The antenna is rotated until a null (a position where the signal is weakest) is reached, and the antenna or its pointer indicates the signal's direction. There should be a visual null indicator on the set so that nulls can be seen as well as heard. Traditionally this is a small meter with a needle, but some modern RDFs have a digital indicator that enables more precise tuning.

With the zero- and 180-degree marks on the azimuth scale aligned parallel to the boat's centerline, bearings can be taken relative to the boat's heading. To obtain magnetic bearings, the boat's course must be obtained from the helmsman at the moment the bearing is taken. LOPs and fixes are obtained as they would be with visual bearings, and often in combination with them. An RDF signal is also useful for homing—heading directly toward a beacon until you can see its light. Certain beacons are distance-finding stations that synchronize a horn blast with a radio signal. Since sound moves much more slowly than radio waves,

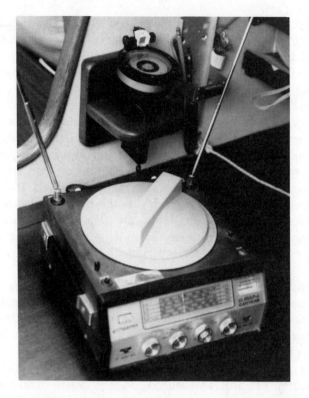

Kelpie's *portable* RDF *on her chart table. Behind the* RDF *is her hand-bearing compass. The prism on top shows her heading from the quarterberth.*

the time difference between the two signals will indicate your distance off the beacon. A rough way of figuring this is to divide the time difference in seconds by 5.5 for a distance in nautical miles.

Helpful as it is, RDF is by no means a precise system of navigation. Inaccuracies are caused by atmospheric conditions, distance from the transmitter, and even deviation similar to the kind that affects a compass. Bearings from dead ahead or astern usually are more accurate, and deviation errors can be minimized by keeping the radio away from metal or magnetic objects. Sometimes error can be caused by rigging or wire lifelines, in which case accuracy can be improved by using rope lashings on the lifelines to break their metallic continuity. When operating an RDF, you have to be careful about recognizing the side from which the radio signal is coming, since there will be two nulls 180 degrees apart. Usually the direction will be obvious, but not always. Some RDF sets have sense antennas to help avoid this ambiguity. When this feature is lacking, there is often no recourse but to turn your boat beam to the signal and temporarily proceed on a new course until a definite change in bearing is noted. Then it should be apparent on which side the beacon is located.

Despite its limitations, I think an RDF should be aboard every long-distance cruiser. Even if you have Loran or an equivalent system, RDF is a valuable backup.

**Depth Sounder**—Another valuable tool for night navigation is the electronic depth sounder. This instrument uses a hull-mounted transducer, which converts electronic signals into high-frequency, inaudible sound waves. The sound waves are sent downward to bounce off the sea bed, and on their return are picked up again by the transducer. The time of travel is measured and then visually displayed on an indicator or recording device. Ideally, a boat should have an indicator above the chart table with a repeater in the cockpit near the helmsman, but it is often possible to mount a single indicator on a hinge near the companionway so that it can be swung in or out for alternate viewing by the helmsman and navigator. Transducers can be mounted inside the hull in a waterbox or cofferdam to reduce drag, but usually they are more effective on the hull's exterior. The depth sounder can be used to navigate in a crude way by relating a chain of soundings to information on the chart—and this, once again, gives you a way of plotting your position that does not rely exclusively on visual reference points. As in the daytime, however, the depth sounder's most important function is simply to check the water depth when you are approaching a shoal.

## SIGHTS AT DUSK

Every offshore skipper should either know the basics of finding a position from the heavenly bodies or else have a competent celestial navigator on board. This is important because electronic instruments can and do fail at sea as a result of antenna damage, power failure, water saturation during heavy weather, or other reasons. Here are a few brief tips on obtaining sights with a sextant after dark:

- See that your eyes are fully night-adapted.
- Take your sight, measuring the angle between the celestial body and the horizon, at the top of a wave so that you can see the true horizon.
- You may be able to see the horizon at night by looking slightly above or below it with both eyes open.
- At twilight, take your sights just as soon as you can see the stars (or other heavenly bodies) in order to have a clear view of the horizon. If you wait until total darkness, the horizon may become indistinct.
- Dawn sights are usually easier because your eyes are better adapted to finding stars and you merely have to wait for the horizon to become distinct.
- To identify a particular star, preset your sextant to the approximate tabulated angle of the star above the horizon and aim the sextant in the proper direction. Once you find it, make fine adjustments to obtain the precise altitude angle.
- Sextant presettings can be determined from a Rude star finder or tables in HO 249 Volume I, published by the Defense Mapping Agency Hydrographic Center.
- HO 249 is extremely helpful because it lists various stars ideal for shooting in groups of seven, and the tables give their computed altitudes and azimuths (bearings).

- A very simple method of determining latitude in the Northern Hemisphere is with a shot of Polaris. For complete accuracy a few corrections must be made, but the altitude of Polaris is very close to your latitude.
- Some planets are relatively easy to shoot because of their brightness. They are distinguished from the stars by their steadiness, or lack of twinkle. Planets can be positively identified with the Rude star finder or *Nautical Almanac*.
- If shooting a bright planet when the horizon is indistinct, use shades to dim the planet slightly.

A few recommended texts for the study of celestial navigation by small boat sailors are: *Celestial Navigation* by Mary Blewitt (J. DeGraff, 1967), *Celestial Navigation Step by Step* by Warren Norville (Second Edition, International Marine Publishing Company, 1984), and *Celestial for the Cruising Navigator* by Merle B. Turner (Cornell Maritime Press, 1986).

## RULES OF THE ROAD

To avoid collisions and comply with the nautical rules of the road, it is essential that the night sailor understand not only the steering and sailing rules, but also the lighting plans for various vessels. For the American sailor, at least, I recommend a study of the U.S. Coast Guard booklet entitled *Navigation Rules* (Commandant Instruction M16672.2, obtainable from the Coast Guard, Department of Transportation, or Government Printing Office). This booklet gives a side-by-side comparison of the Inland Rules used on U.S. waters and the International Rules used on the high seas. Fortunately, the two sets of rules have finally been brought into close agreement with each other.

On a dark night when you cannot see a vessel's hull, her lights may at first appear confusing, but not so after the basic system for navigation lights is understood. To interpret what you are seeing, you need only recall the fundamentals already presented in Chapter 1. The red (port) and green (starboard) sidelights, with their arcs of visibility running from dead ahead to 22.5 degrees abaft the beam, show which side of the vessel you are seeing and, in a general way, how she is headed. If you see only one sidelight, you are seeing one side of the vessel, but watch out if you see both sidelights, for she is headed directly toward you. Incidentally, you can usually see the red light sooner than the green one. A vessel 50 meters (164 feet) or more in length is required to carry range lights, which show the exact direction in which she is traveling. These are white "masthead" lights on separate masts with the after light mounted at least 2 meters (6.5 feet) higher

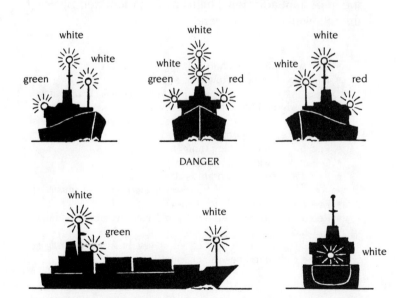

*Navigation lights for a power-driven vessel 164 feet or longer. Notice how the range lights, appearing closer together or farther apart, indicate the vessel's heading.*

than the forward one. Each mast light has a visibility sector of 225 degrees, with the blacked-out sector facing aft. When these two mast lights are close together (appearing to be one nearly over the other) and a sidelight is visible, the ship is headed toward you, and evasive action may be necessary to avoid a collision. When appearing to be spread far apart, however, the range lights show that the ship is traveling perpendicular (or nearly so) to your line of sight. If the range lights appear to be blacked out, but you see a white sternlight, the ship is headed away from you.

What has been described is the basic system that always applies, but bear in mind that some vessels show many additional lights. For example, a tug towing or pushing may resemble a Christmas tree, while passenger vessels or ships at anchor are often lit up like Times Square. Many of the important additional lights can be identified by using the following bit of doggerel:

- One white light—anchored at night.
- White over white—a tow is in sight.
  *(pushing or towing alongside or close astern)*
- Three whites in a row—barge is in tow.
  *(when tow length exceeds 200 meters)*
- White over red—pilot boat ahead.
- Red over white—fishing tonight.
- Green over white—trawling tonight.
- Red over red—the captain is dead.
  *(vessel not under command or aground)*
- Red over green—sails to be seen.
- Red over white over red—restricted maneuverability astern or ahead.
  *(also for diving and dredging)*
- Three reds in a line—the channel is mine.
  *(vessel constricted by her draft)*

For complete lighting details and exact meanings, consult the Coast Guard's *Navigation Rules.*

When converging with another vessel, you can determine whether a danger of collision exists by applying the principle of change in bearings. Take a bearing on the vessel, and a short time later take another. If there is no change in the bearing angle, you are on a collision course. You can breathe easier if the bearings change, however, for this indicates that the converging vessel will pass either ahead of or behind you. In the latter case, be sure the angle changes sufficiently to put her very far behind you. This is particularly true of ships, which cannot easily maneuver or slow down. In converging situations, always take pronounced evasive action well ahead of time so that there can be no confusion about your intentions.

As a matter of fact, it is the safest policy at night to give any approaching vessel a wide berth. If you are close, shine a strong light on your sail to be sure you are seen. In coastal waters it is best to stay out of shipping lanes or channels if at all possible. Sometimes you can skirt along the edge of a channel, just outside of it, where a ship cannot come close to you without running aground. If you need to venture into waters used by ships, carry a good radar reflector on the mast or aloft on the backstay to assure greater visibility on the screen of a vessel using radar.

For night races, standard racing rules are sometimes modified in accordance with the United States Yacht Racing Union Rule 3.2(b) (XXVIII). In the Chesapeake Bay, for instance, the following regulations exist from sunset to sunrise: "When two yachts are on the same tack within three overall lengths of the larger yacht, the yacht being overtaken shall maintain her proper course. The overtaking yacht shall keep clear, and neither yacht shall bear away toward nor luff the other." The spirit of prudence implicit in this modification is a wise guideline for all nighttime navigation.

# 5

# Safety
# Considerations

Important anytime, safety becomes a very special consideration at night. Darkness not only impairs the vision, but can cause confusion and disorientation, which may lead to mistakes and perhaps even dangerous problems. Many of us exhibit a somewhat carefree complacency in the daytime that seldom causes any harm, but we definitely need a more cautious attitude after the sun goes down.

## STAYING ON BOARD

Bear in mind that anyone who should happen to fall overboard will be very hard to find in the dark. How well I remember hearing, when I was serving on the Chesapeake Bay Yacht Racing Association's safety committee, how a fellow member of the committee lost his foredeck man overboard during a night race. The accident occured on a seaworthy boat in semiprotected waters not far from shore, yet the man could not be recovered and drowned. Apparently the boat had been struck by a sudden squall, and the foredeck man went overboard while lowering the jib. Even though there had been a number of lightning flashes to give warning of the approaching storm, its severity had been hard to judge in the darkness, and all the competitive racers had tried to carry sail until the last possible minute before the blow. The incident is a grim reminder that accidents can indeed happen anywhere, even on seaworthy boats and to highly experienced sailors.

Recovering a person who has fallen overboard is a problem we'll soon discuss, but first of all, let's think about how we can best keep that person on board. As mentioned earlier, we must guard against hazards such as open hatches that can be stepped into or cushions, lines, and sails that can cause tripping or slipping. Keeping the boat tidy and shipshape will go far in eliminating booby traps.

Adequate handholds are essential to keeping your crew on board. Any gaps on deck where there is nothing to grab should be fitted with handgrips. In most cases these can be lifelines, pulpits, or grabrails. If there is a gap between the bow pulpit and the forward ends of the lifelines for a low-cut jib to pass through, there should be a temporary line, fitted with a snapshackle perhaps, to close the gap when sailing in rough waters or at night. Sometimes this gap can be filled sufficiently by crisscrossing two lines between the forwardmost lifeline stanchion and the pulpit's after end. The crisscross pattern allows the jib foot to clear, but still gives a kneeling crewmember something to grip.

Handrails are needed along most of the cabin trunk. Quite often the below- and abovedeck rails can be bolted together, one above the other, along the edge of the cabintop. This is a secure arrangement that minimizes leaks and affords good gripping below as well as on deck. Another spot where handrails are needed is near the companionway. By all means check that handgrips are adequately bolted and have large washers. Poorly fastened grips pull off. Hard as it is to believe, I was told by the Coast Guard of one stock boat whose handrails were not fastened at all, but merely glued on! Of course, some grips are used only to achieve balance, but assume there will come a time when a wave will knock you off balance and you'll snatch at the grip, putting more than your full weight on it.

The safest policy for night sailing is to have double lifelines. The top line should be well above knee level (at least 24 inches high) and should be white vinyl-coated or even jacketed in yellow plastic for visibility. I like to lash the lifelines to the pulpit, primarily because the lashings can be cut, which may be helpful in recovering a person overboard. Pelican hooks also allow dropping the lifelines, but they can open inadvertently unless taped or lashed.

Be sure there are no slippery surfaces that could cause a fall. The greatest hazards are varnished surfaces wet with rain or spray. Varnished hatch covers, bowsprits, companionway steps, etc., should be covered with strips or mats of nonskid material, available in handy pressure-sensitive adhesive form and in both dark and light colors to contrast with the surfaces to which they are applied. An alternative method is to paint slippery surfaces with a nonskid coating that contains pumice. Every crewmember should wear proper nonslip deck shoes.

Toerails are often varnished, and if so, they too should be skid-proofed as much as possible. Consider painting the covering board (the most outboard deck plank) white so that it contrasts with the darker, varnished rail and helps define the deck edge at night—a good tactic for dark fiberglass or plywood decks as well. White rails may help define a vessel's edge in calm conditions, but there are times when a white rail will blend with the foam at the waterline and afford little if any contrast.

The surest way to keep a person on board is to attach him or her to the boat with a tether, a short safety line,

*The author sitting at the helm, garbed in a flotation jacket with safety harness and tether. Tethers are essential on passages at night when there is only one person on deck standing watch or steering.*

connected to a safety harness. The basic harness is a belt with shoulder straps, although there are also more elaborate types that provide some buoyancy with inflatable collars. A safety line should have a large interlocking snaphook (such as a carbine or antitripping Gibb type) at its end so that the person wearing the line can snap onto a strong point on the boat using one hand. Some sailors prefer snaphooks at both ends of the safety line so that the wearer can easily release himself if held underwater. Strong points to which safety lines are snapped can be some secure part of the rigging, stanchion base, handrail, or other solid fittings, or special through-bolted eye straps can be installed. Snapping onto lifelines is not recommended, as some cannot accept the shock loading of a heavy crewmember falling overboard or across the boat and coming to an abrupt stop at the end of his tether.

In rough weather crewmembers moving along the deck should be sure that anytime they must unhook to move they have a firm grip on a rail with one hand. Offshore or in very rough conditions it is often a good idea to rig cable travelers or jacklines running from the cockpit to foredeck on each side of the boat to permit uninterrupted travel along the deck. These travelers should be rigged well inboard, as close as possible to the boat's centerline. Ronstan Marine makes a latchway system consisting of a small cylindrical turnstile (accepting a safety line's hook) that allows movement along a traveler cable without the need to unhook and rehook when passing an obstruction such as a stanchion.

Tethers are essential on passages at night when there is only one person on deck standing watch or steering. On *Kelpie* we have a rule that when one person is on deck at night, he or she must remain tethered in the cockpit until another person comes on deck. This may seem overly cautious, but I know of several people lost overboard when left alone on deck. In one case, two crewmembers were on watch, and one merely stepped below for a moment to light his pipe. When he returned to the cockpit, the other had disappeared, never to be found. Perhaps he had gone to the afterdeck to relieve himself and had slipped and fallen overboard. No one will ever know.

Another important safety measure when singlehanding or standing lone watches at night is to tow a long floating line that has been rigged to trip any self-steering device that may be in use. The tripping mechanism need be no more elaborate than a snapshackle that is released by a tug on the floating line. Should the singlehander fall overboard, he will hopefully be able to grab the towed line and hang on, or at least disengage the self-steering. The

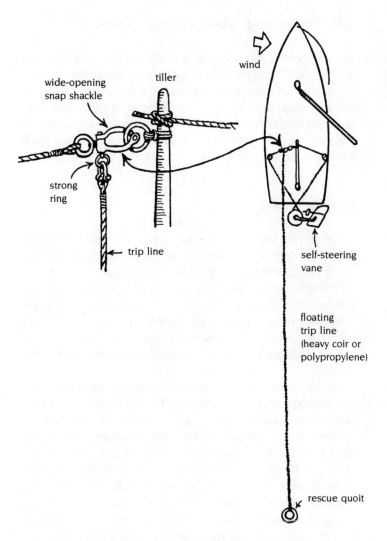

wind

wide-opening
snap shackle

tiller

strong
ring

trip line

self-steering
vane

floating
trip line
(heavy coir or
polypropylene)

rescue quoit

*Tripping the self-steering vane.*

boat should be trimmed to carry a slight weather helm so
that she will round up into the wind when the self-steering
is tripped.

65

## MAN OVERBOARD GEAR AND PROCEDURES

Let's suppose that, despite our efforts to keep the crew aboard, someone goes over the side. We must get back to the person in the quickest possible time, especially if the water is cold and if he is not wearing a PFD (personal flotation device) or any garment that supplies buoyancy. In addition to the problem of finding the victim in the dark, we also face the difficulty of getting him back on board.

### Basic Principles

There can be few hard and fast rules for recovery because circumstances vary according to the handling characteristics of your boat, the sea conditions, the sails carried, and your crew's familiarity with rescue methods. Still, several basic principles should be kept firmly in mind. First of all, the crew must be alerted at once when a person is seen falling overboard, and *simultaneously* close-at-hand flotation such as seat cushions should be thrown close to windward of the victim.

Assign someone the specific duty of watching and pointing to the victim. At night or in seas a man-overboard pole should be dropped over the side, attached to a horseshoe buoy and strobe light (more information on this will follow shortly). The vessel should not be allowed to get far from the victim, especially in conditions of poor visibility. If it is necessary to stay on course for even a moment, or if the person is not seen falling overboard, the compass heading must be noted, so that the reciprocal course can be used to return to the victim. In most circumstances, however, it is better to turn immediately and heave-to close to the person overboard.

The engine can be used to increase maneuverability, but it must never be started when there are lines overboard. It should be well muffled, and the propeller must be kept well clear of the person overboard. Ensuring against injury from the prop may require stopping the engine and drifting down on the victim. There is some disagreement as to whether the victim should be approached from the windward or leeward side, but in my opinion the windward approach is usually better because the vessel will drift toward the victim rather than away from him. A windward approach also means that recovery will take place from the boat's leeward side, which is easier because it is normally lower than the windward side and the boat forms a protective lee. The risk in a windward approach is that the vessel will drift over the victim and force him down, but this involves little danger if the victim is wearing a PFD, and if there is ample crew at the leeward rail to hoist him aboard.

## Rescue Lights and Other Aids

Adequate lights are a great aid in night rescues. For any extensive or regular night passage work, I would recommend that each crewmember carry a personal high-intensity strobe light, and that the boat be equipped with at least one of the larger, floating kind that can be attached to horseshoe or ring buoys. Personal strobes come in small sizes that can be carried in one's pocket or strapped to an arm or a PFD. Less effective but cheaper and far better than nothing are incandescent personal lights or even Cyalume light sticks that can be bent to produce a nonvolatile green or white glow.

Floating strobe flashers are usually attached to a horseshoe buoy and man-overboard pole with a tether. The typical man-overboard pole is 11½ to 15 feet long with an integral float about eight feet from its top and ballast at its bottom enabling the pole to float upright. It has a red and yellow flag at its top and is most often mounted on the stern pulpit in such a way that it can be quickly thrown overboard. Some poles have strobe lights at their tops, and these are very effective in high seas.

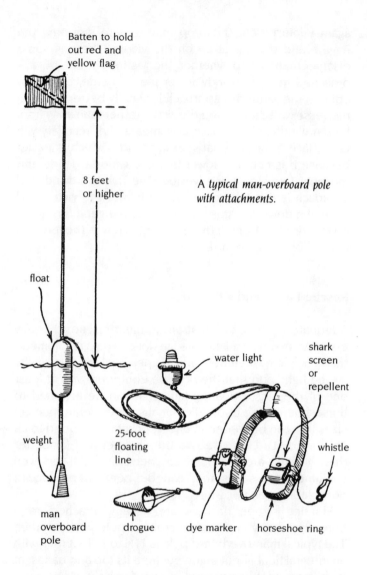

Batten to hold out red and yellow flag

8 feet or higher

*A typical man-overboard pole with attachments.*

float

water light

shark screen or repellent

weight

25-foot floating line

whistle

man overboard pole

drogue

dye marker

horseshoe ring

Most man-overboard strobe lights attached to horse-shoe buoys are hung upside down on deck and are ballasted to self-right when thrown overboard. The strobe is automatically activated when the light self-rights. I have had trouble with water leaking inside the rubber cap of a well-known make of man-overboard light when it was hung upside down. Although the problem was overcome by putting a hose clamp around the rim of the cap, I will never again use the rubber-cap type.

Needless to say, there should be a powerful searchlight—preferably 200,000 candle power—to look for the victim. Powerful lights that plug into the ship's power are easily obtainable, but if possible, I would try to find a light that has its own batteries, since this obviates the need for a plug/receptacle and a cumbersome electric cord that restricts carrying the light any distance from its receptacle. Automotive stores sometimes sell such lights, but they are not designed for the marine environment and can corrode relatively soon. Still, I prefer them for their handiness and portability. Whoever operates the searchlight must be extremely careful not to shine it in the eyes of the helmsman and other crewmembers.

The visibility of life buoys, foul-weather gear, PFDs, and the like is enhanced with the use of reflective tape such as 3M Scotch Lite or the retro-reflective tape composed of vast numbers of tiny glass beads, which (unlike a mirror) always reflects light shone on it back to the light source.

Other items that may aid in finding a person overboard are loud referee-type whistles, dye marker, and white parachute flares. Worth special mention are the inflatable, helium-filled balloons that can be attached to PFDs and sent aloft by the victim (a well-known make is Life-A-Line by Rossam Industries in Florida). Tethered from the end of a long line, an airborn balloon is easily spotted above high seas. For effectiveness in the dark, balloons are covered with reflective material that can be picked up by spotlights or radar.

Dye marker can be a real help if the person overboard releases it at once, since the brilliant yellow stain can usually be seen even without light from close by. Although the dye may spread some distance from the victim, it tells the rescuers that they are in the right ballpark. Whistles can be very effective if the rescuers are downwind of the victim and the engine is not running or is sufficiently muffled. A whistle and small pack of dye are usually part of the equipment attached to a horseshoe buoy (along with a man-overboard pole, small drogue, and strobe light), but it is also a good idea for an on-watch crewmember to carry a dye pack and whistle in a pocket in case he or she falls overboard and cannot reach the horseshoe with its attachments.

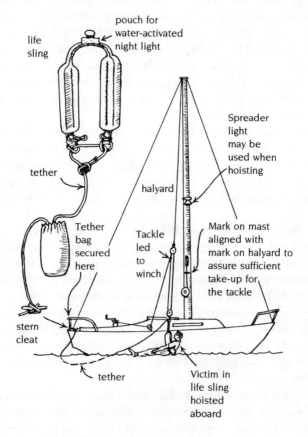

pouch for
water-activated
night light

life
sling

Spreader
light
may be
used when
hoisting

tether

halyard

Tether
bag
secured
here

Tackle
led
to
winch

Mark on mast
aligned with
mark on halyard to
assure sufficient
take-up for
the tackle

stern
cleat

tether

Victim in
life sling
hoisted
aboard

*The Sailing Foundation's Lifesling system.*

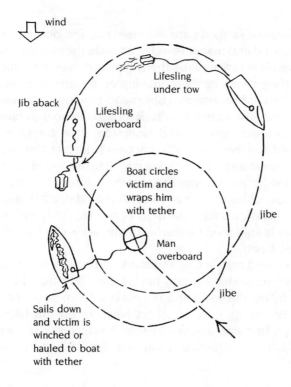

wind

Lifesling
under tow

Jib aback

Lifesling
overboard

Boat circles
victim and
wraps him
with tether

jibe

Man
overboard

jibe

Sails down
and victim is
winched or
hauled to boat
with tether

## Lifesling

In current vogue for rescuing a victim overboard is the Lifesling (sometimes called Seattle sling), developed by the Sailing Foundation of Seattle, Washington. This device consists of a buoyant flexible horsecollar (similar to the hoisting slings used by helicopters) attached to the boat by a long length (perhaps 150 feet) of floating polypropylene line. A water-activated light, which fits into a pouch on top of the sling, increases suitability for use at night.

The recommended method of use in most circumstances is as follows: Immediately after the person is seen falling overboard, the boat is luffed head-to-wind, then slightly beyond with the jib aback. The Lifesling, normally mounted on the stern pulpit, is thrown overboard. Its tether, which has been coiled and stuffed in a bag, pays out as the boat falls off onto the new tack. The boat is turned downwind and maneuvered to encircle the victim with the tether. The victim grabs the tether and uses it to draw in the sling, which he then puts under his arms. Meanwhile, the boat has once again come head-to-wind, and sails are dropped. The victim is pulled close to the boat by hand or with a winch and secured alongside. If he cannot climb aboard and there is insufficient crew to lift him, then he can be hoisted aboard by a tackle rigged to the end of the main halyard. I would recommend that the halyard be premarked to show when its end is 10 to 15 feet above the deck in order to allow ample take-up of the tackle. If additional power is still needed, the fall of the tackle can be led to a sheet winch.

I am told that in England there is a feeling that circling the victim under sail may not be easy, and that the sling may follow the yacht and not make a loop. Nevertheless, the Seattle group and others have tested the Lifesling system in numerous practice sessions, and they claim it works. It seems particularly appropriate for a shorthanded boat.


## Other Returning Methods

Similar to the Lifesling method, at least in the initial phase, is the so-called quick-stop method presently in use at the U.S. Naval Academy. Way is rapidly lost by luffing head-to-wind and then momentarily heaving-to with the jib aback. After most headway is lost, the boat is turned downwind, the headsail dropped, and the mainsail centered. When the victim is abaft the beam, the boat is jibed, and he is approached on a course of 45 to 60 degrees off the wind. A line is thrown, the boat is slowed by pushing or pulling the main boom (with a line led forward) against the wind, and the victim is hauled aboard by the crew. With practice (and an ample crew) this system works well.

A traditional method taught by sailing schools, especially

in England, is the "reach-tack-reach" system, where the yacht sails away from the victim on an apparent beam reach, then comes about and reaches back on an almost reciprocal course. The approach is from just to leeward of the victim so that sheets can be released to slow the boat before contact is made. At the Yachtmaster Instructor Conference in England (November 1984), at which man overboard procedures were practiced, the reach-tack-reach system continued to be favored over the quick-stop method. This suggests that sailors generally prefer the most familiar system and that any procedures must be practiced for greatest success. Quick-stopping, it was said, sacrifices control and maneuverability of the boat, but I prefer that the boat stay close to the victim, especially at night.

Another traditional approach, which keeps the boat close to the victim yet without giving up steering control, is the quick jibe. As soon as a person goes overboard, the rudder is turned hard to leeward; the boat jibes and continues to turn until she is nearly head-to-wind next to the victim. This is the fastest method of returning, but obviously the boat must be rigged to sustain an immediate flying jibe without damage. This means no running backstays or sails that must be lowered before changing tacks. Also, of course, the weather conditions must be such that jibing causes no damage. Even during heavy weather, however, a controlled jibe, with sheets of boomed sails pulled in, might sometimes be better than trying to come about against a steep sea that could put the boat in stays. This will depend to a large extent on the handling characteristics of your boat and the sea conditions.

Thorough familiarization with your boat's turning radius and other handling characteristics, together with a lot of daytime practice, is needed to avoid any disorientation caused by jibing at night. One way to minimize confusion when beating is to glance at the windward transverse (side) lubber line on your compass just before jibing so that you can figure on establishing an approximate 90-degree heading to your original course after jibing. As I said before, someone should be watching and pointing at the victim during the maneuver. If the jibe can be made at once, the victim can often be kept in view even if it is quite dark.

Object overboard drills can be a useful way to familiarize yourself with turning maneuvers and retrieval techniques. It should be realized, however, that picking up an object is no substitute for retrieving a person. A lightweight, shallow-draft object, such as a floating seat cushion, will drift much faster than a human and is therefore not a very realistic simulation. A better idea is to use a coconut, which more nearly approximates the drift of a person and has a shape closer to that of a human head. Practice sessions involving the retrieval of an actual person should only be done with ample crew on board except in extremely favorable weather conditions. Obviously, the person overboard should wear a PFD, the drill should not be practiced in cold water, and there should be an assortment of boarding aids.

## Boarding Systems

Half the problem of recovery is getting an exhausted or unconscious victim back on board when the boat is shorthanded. Using a tackle rigged from the main halyard as previously described is one way to do it, but on some boats it may be simpler to use the vang tackle rigged from the main boom. Quite often the vang tackle is right at hand, and the boom can be used to swing the victim aboard. This method may require dropping the lifelines, but doing so should present no serious problems if the lifelines are secured with pelican hooks or lashings that can be cut.

If there is no life sling, tie a large loop in the end of a line for the victim to put under his arms. A rather neat trick is to tie a bowline on the bight so that there are two loops, one to fit under the victim's arms, the other under his seat or

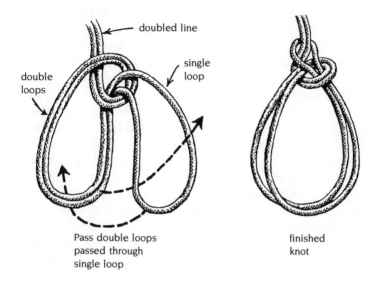

doubled line

double
loops

single
loop

Pass double loops
passed through
single loop

finished
knot

*Bowline on the bight.*

bent legs. To tie the bowline on the bight, turn the rope back onto itself and tie as shown in the accompanying illustration. The looped line should be thrown to the victim as soon as he is close enough to grab it; then he climbs into the loop or loops and is pulled alongside. If the victim is unconscious or disabled, it may be necessary to send a swimmer into the water to assist. That person must have on a PFD and be tethered with a line.

The U.S. Naval Academy has devised a "throwing sock," which contains a coiled heaving line that automatically pays out when thrown. A similar device, called the Omega Rescue Throw Bag, is now sold by West Marine Products. Another useful device is the Hambly Lifeloop, a PVC-covered wire loop at the end of a hollow plastic tube, available from Henderson (no relative of mine) Pumps in Ireland. The Lifeloop can either be thrown to the victim or used to snare him when alongside.

Other means of helping a conscious victim aboard are

with a ladder, a scramble net, or a parbuckle (a sling with one end fixed and the other lifted to hoist an object on board). Many modern boats are fitted with permanent stern ladders that fold down into the water. These are a good idea, but care must be taken to see that the victim is not hit on the head by the counter of the boat when it is pitching in heavy seas. A substitute for a permanent ladder is one of rope that can be rolled up and fastened to the stern pulpit or lifeline with "rotten" twine that can be easily broken. A pull-down line from the rolled-up ladder can be left hanging overboard so that it can be reached even if a singlehander goes overboard. Such a ladder should have a weighted lower rung to make it sink. The "stirrup ladder," a kind of Jacob's ladder (a ship's ladder consisting of rope or chain sides with wood or metal rungs), has been highly recommended. Beware of portable swimming ladders that can come loose and possibly injure the victim in rough seas. Even if you have no satisfactory ladder, a loop of line hanging down can make a good foothold for the victim.

A boarding device favored by many British yachtsmen and endorsed by the Royal National Lifeboat Institute is the scramble net, a sort of miniature cargo net hung over the side. Such a net not only facilitates climbing aboard, but also allows crewmembers to climb down and assist the victim. The maker of Neptune Nets in Diss, Norfolk, England, has several practical suggestions about their use: the net should be secured at two corners with red lashings (for easy identification) to preselected strong points on the deck near the rail (perhaps stanchion bases) and then thrown over the top of the lifelines. The unsecured corners should be weighted (with old shackles, perhaps) to make the net sink quickly and ride two or three feet below the surface. The net is normally used on the leeward side. Two moderately long ropes are attached to the lower corners and led up to the deck so that the net can be used as a parbuckle if the victim is completely exhausted or unconscious.

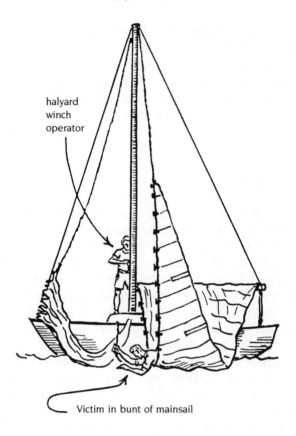

halyard
winch
operator

Victim in bunt of mainsail

*Using the mainsail as a parbuckle.*

Sails can also be used as parbuckles to haul incapacitated victims aboard. Drop the mainsail, detaching its luff
rope or slides, then lower the bunt over the side. The
victim can climb or be scooped into the bunt; hoisting
aboard is accomplished by winching the sail's head aloft
with the halyard. The U.S. Naval Academy has devised a
drill that uses a small genoa jib as a parbuckle. The clew
and tack are shackled together, while the head is tended
aft; then the foot, which forms a large bight, is lowered over
the side and used as a scoop.

If you have a rubber raft on board that is carried partially inflated (or can be quickly inflated by $CO_2$), it can assist in getting an injured person aboard, but I would not spend a lot of time pumping up a rubber dinghy. The sooner the victim can be gotten back on board the better; in cold waters, especially, I would prefer a quicker parbuckle method.

## Search Patterns

Even if the victim is not seen going over the side and no one knows when the accident occured, don't despair. I know of several cases where victims were overboard for hours but were eventually recovered after the boats were turned onto reciprocal courses. In the event that the victim is not found by following a reciprocal course, a systematic search pattern should be instituted. As soon as the victim is discovered missing, a boat under power should usually make what is known in the U.S. Navy as a Williamson turn. The boat is turned 60 degrees away from her heading and then brought around the other way to put her back on her track on a reciprocal heading. The victim will drift to leeward, but the boat will also make leeway, and there may not be very much difference between the leeway and drift. Contrary to what is sometimes written, set from the current will seldom be a factor because both victim and boat are normally subject to similar conditions. If the two are in different conditions, of course, the difference must be considered.

A systematic search pattern consists of back-and-forth sweeps some distance apart, but obviously, no farther apart than a person can see. Be sure that search patterns are not initiated until it is certain that the returning boat has gone well beyond the point where, it is "guesstimated," the victim fell overboard. If you are under sail, it is usually best to begin a search pattern to leeward of the victim, partly because the searchers can then hear his shouts or

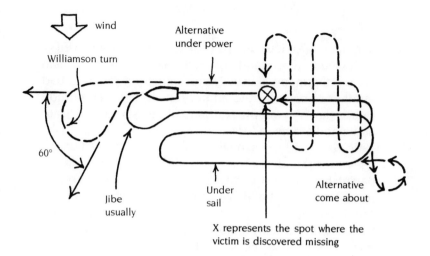

wind

Williamson turn

Alternative under power

60°

Jibe usually

Under sail

Alternative come about

X represents the spot where the victim is discovered missing

*Systematic search patterns.*

whistle. The boat reaches back and forth and tacks at the end of each sweep. If you are under power with sails down, a leeward search is in some cases still a sensible approach, but the victim's sounds are not as likely to be heard above the noise of the engine, and a boat without sail will tend to roll considerably in beam seas, creating a condition that could hamper the efficiency of the search. The most important matter, however, is to work systematically so that no area of the sea surface is missed. If the victim is not immediately found and you are in populated water, use your radio or flares and strobe lights to contact the Coast Guard or nearby boats to help with the search.

## JUDGING THE WEATHER

Weather is a bit more difficult to judge at night, since the sky cannot easily be seen. You can tell when the night is clouding over, however, by observing the clarity of celestial bodies. Although it is often difficult to tell precisely what kind of clouds are forming, you can usually gain enough information to determine a course of action. A thin veil of high clouds that merely dims the stars or forms a halo around the moon is usually only a first indication of bad weather to come in the future, but thick, lower-altitude clouds that blot out the stars and fill the sky may indicate an immediate need for caution.

Being an old Chesapeake Bay sailor, I have a healthy regard for thunderstorms. Particularly treacherous in my part of the world but worthy of respect wherever they occur, these storms can build fast and strike with short notice in conditions of intense heat and humidity. In the middle latitudes they usually build to the west of you (particularly if there is land to the west), but be alert as well when thunderstorms are close by to the north or south—I have seen a few humdingers come from these directions. In the late afternoon you may see lofty vertical clouds rising in the west, nor'west, or sou'west; this is a tip-off to be careful.

The most noticeable warning of a thunderstorm at night comes from the lightning and thunder it produces. Since sound travels a great deal more slowly than light, the storm's distance can be judged by timing the interval from a lightning flash to the subsequent sound of thunder. Count the number of seconds between the flash and thunderclap and divide the number by five to get an approximation of the number of miles. Another rough way to judge the closeness of the storm is to turn on the radio and hear the amount of static produced.

Thunderstorms of the isolated air mass type can be vicious, but they seldom last long. Frontal storms and squall lines that are the forerunners of cold fronts are often more dangerous, although they usually can be predicted by studying a weather map (from a newspaper or TV) before getting underway. Radio broadcasts such as those on VHF/FM 162.55, 162.40, and 163.275 MHz, which are updated every few hours, also provide warnings of approaching fronts or stormy lows, and the listener is often cautioned when conditions are ripe for isolated thunderstorms. But although weather forecasts are extremely useful, the seasoned sailor never considers them gospel. They are often wrong in the matter of timing, and bad weather can arrive much sooner or later than predicted.

If you are out sailing at night and a thunderstorm seems imminent, the simplest action is usually to lower your sails and start the engine. Stop your sails well, doublecheck your position, and figure a safe compass course that provides searoom so that you will not run aground or collide with anything when the rain further reduces visibility. Put on flotation gear and perhaps safety harnesses, and see that ground tackle is ready in case you have to anchor. Make sure that all crewmembers stay away from rigging or large metal objects that could conduct an electrical charge in the event the mast is struck by lightning. If the seas get rough, power slowly into them with just enough speed to keep the bow from blowing off. When you have no engine and there is plenty of room to leeward, you might run off under bare poles. A typical air mass thunderstorm probably will last less than an hour, so you can simply wait it out and then power back to your harbor or continue with your sail.

When making a cruising passage offshore, it is always the safest policy to reduce sail at night if the weather seems at all threatening. Of course, fully crewed racers will carry all possible sail and shorten down as the wind increases, but it is prudent for a shorthanded cruiser to reduce sail early. In squally weather at sea when conditions are unstable, there could be the possibility of a twister (mini-tornado) or microburst, a low-level wind shear that generates a very strong downdraft and outflow over a concentrated area.

How well I remember being hit by a small twister at night in the Gulf Stream when sailing *Kelpie* from Bermuda to the Chesapeake a few years ago. The weather was squally, and we decided to shorten down to a number two working jib. Although I felt ultraconservative (almost chicken, in fact), we were proceeding on course, making a steady four knots, and were quite comfortable despite some steep quarter-

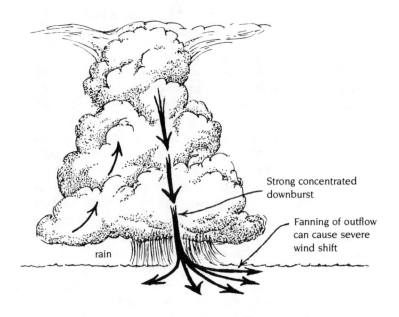

Strong concentrated downburst

Fanning of outflow can cause severe wind shift

rain

*A microburst (thundercloud is not always present).*

ing seas. Leaving my son Rip at the helm, I went below for a nap. As soon as I had closed my eyes, I heard Rip exclaim, "Oh, my God!" An instant later *Kelpie* was knocked down, and there was a great commotion on deck with splashing water and rattling gear. When she righted, I rushed topside to see what had happened. Rip was OK and said that he had glanced to weather in time to see a rapidly approaching apparition in the form of a small waterspout, then about fifty feet away. He ducked his head and grabbed the wheel with both hands, but felt himself lifted bodily off his seat as the twister swept across the cockpit. He was mighty thankful to have been wearing his safety harness with a stout lanyard secured to the binnacle. The episode lasted only a few seconds, and *Kelpie* was unscathed, but we learned the value of being cautious. In any potentially threatening situation when visibility is limited, shorten down early, use safety lines, have your gear and sails well secured, and keep all vulnerable ports or hatches closed.

**APPROXIMATE BLOOD ALCOHOL PERCENTAGE**

| Drinks | Body Weight in Pounds | | | | | | | | |
|---|---|---|---|---|---|---|---|---|---|
| | 100 | 120 | 140 | 160 | 180 | 200 | 220 | 240 | |
| 1 | .04 | .03 | .03 | .02 | .02 | .02 | .02 | .02 | BE |
| 2 | .08 | .06 | .05 | .05 | .04 | .04 | .03 | .03 | CAREFUL |
| 3 | .11 | .09 | .08 | .07 | .06 | .06 | .05 | .05 | |
| 4 | .15 | .12 | .11 | .09 | .08 | .08 | .07 | .06 | OPERATION |
| 5 | .19 | .16 | .13 | .12 | .11 | .09 | .09 | .08 | IMPAIRED |
| 6 | .23 | .19 | .16 | .14 | .13 | .11 | .10 | .09 | |
| 7 | .26 | .22 | .19 | .16 | .15 | .13 | .12 | .11 | |
| 8 | .30 | .25 | .21 | .19 | .17 | .15 | .14 | .13 | DO NOT |
| 9 | .34 | .28 | .24 | .21 | .19 | .17 | .15 | .14 | OPERATE |
| 10 | .38 | .31 | .27 | .23 | .21 | .19 | .17 | .16 | |

One drink is 1¼ oz. of 80 proof liquor, 12 oz. of beer, or 4 oz. of table wine.

*The effect of alcohol on boat operation. (Courtesy U.S. Coast Guard)*

## GROG RATIONS

Much as I hate to say it, there should be a limit on the amount of alcohol consumed by the crew and skipper, particularly at night. The operation of a boat, as well as an automobile, is adversely affected by imbibing excessively; in fact, the U.S. Coast Guard estimates that alcohol may contribute to half of all boating accidents.

Just what constitutes an excessive amount of liquor is open to some debate. It depends on such matters as individual metabolism, body weight, how much the drink is diluted, the proof of the liquor, the amount of time between drinks, the amount of food recently eaten, and even fatigue and exposure to heat. The accompanying chart provides some rough guidelines, and needs to be taken seriously. Especially at night, when it is easy to become confused, you need all your wits about you.

# 6

# Passage Planning

## USEFUL PUBLICATIONS

For any kind of passage that involves sailing at night, considerable planning and preparation is necessary well ahead of time. To begin with, buy an up-to-date cruising guide of the area you intend visiting, and study it carefully. Available from marine publishers, book stores, and chandlers, a proper cruising guide will help you select the most suitable ports of call by providing such vital information as accessibility and protection of harbors, navigation details, facilities, sources of supplies, tips on local weather and tides, and points of particular interest. The guide will also familiarize you with the harbor you are planning to visit and help you determine the feasibility of entering it after dark. In some cases, entering at night may not be advisable because of a treacherous inlet, a paucity of lighted navigation aids, a confusing background of shore lights, a profusion of fish traps, or some other reason.

The next step in planning a passage is to acquire proper

charts and *Light Lists* (already mentioned in Chapter 4) so that you can plot your course with an eye to helpful lights and sound signals en route. You should lay out your course between well-lighted navigation aids and pick out radio-beacons and distant lights from which bearings can be taken along the way, as discussed in Chapter 4. Try to estimate your speed with various probable winds or under power in both fair and adverse current, in order to figure your progress and thus where you will most likely be at night. Be sure to select alternative anchorages in which you can seek shelter in unfavorable weather.

As mentioned previously, it may pay to leave your home port in the dark so as to arrive in difficult or unfamiliar waters during daylight. Sometimes it is helpful to plan your arrival just before sunrise so that you can take advantage of navigation lights during the passage itself, then heave-to until dawn in order to enter a difficult anchorage during daylight.

Tides and currents are major considerations in some harbors and passages; for this reason it is advisable to have up-to-date tide and tidal current tables for the area you are visiting. Another valuable publication is *Notice to Mariners* (published weekly by the Coast Guard) which announces recent changes to navigation aids and thus comprises a vital supplement to your charts and *Light Lists*. A few years ago a yacht was wrecked on the reefs off Bermuda because the radiobeacon on which the boat was homing had been moved. The accident could have been avoided by consulting a recent *Notice to Mariners*.

For long-distance passages, you'll want other publications as well, such as *Nautical Almanac* (published by the U.S. Naval Observatory), *Radio Navigational Aids* (published by the Defense Mapping Agency Hydrographic/Topographic Center), *Sailing Directions* (published by the U.S. Naval Oceanographic Office), *Worldwide Marine Weather Broadcasts* (published by the U.S. Department of Commerce), and U.S. *Coast Pilots* (published by the National Ocean Survey) or the British Admiralty Pilots.

Pilot charts are particularly useful in the planning stage, for they give important information in percentage form about gales, calms, wind directions and strength, current, fog, ice limits, atmospheric pressure, magnetic variation, mean atmospheric pressure, and so forth. This knowledge is vital in plotting the route and choosing the proper time of year for your passage. You should keep firmly in mind, however, that pilot charts cannot predict; they merely give percentages based on past records. The charts are published quarterly by the Defense Mapping Agency Hydrographic/Topographic Center, and are available from local chart outlets or from the DMA Office of Distribution Services (DDCP), 6500 Brookes Lane, Washington, D.C. 20315. Another useful resource for route planning is *Ocean Passages for the World*, published by the British Hydrographer of the Navy and available from agents for the sale of Admiralty charts. For planning your ports of call, you may want to consult the *World Port Index*, published by the U.S. Naval Oceanographic Office.

## ADDITIONAL GEAR NEEDED

Boat preparation and safety gear for night sailing has already been discussed, but there should be extra equipment when a lengthy passage is planned. A voyage or offshore passage may require self-sufficiency to a very high degree; therefore, the boat must be equipped with spares of almost everything. She should carry a complete set of tools; repair kits for plumbing, electrical, engine, sail, and rigging; a completely equipped liferaft or rafts; heavy-weather gear such as storm sails, rodes, sea anchor, extra pumps, and storm slides for the companionway; extra food, water, and fuel; and a complete medical kit with suitable instruction manual. For the offshore boat I would certainly recommend a large spray dodger that unfolds upward to cover the companionway and forward end of the cockpit. This will shelter the on-deck crew and permit good ventilation below, greatly improving the comfort of everyone, especially at night in cold, wet weather. Likewise, weather cloths lashed between the lifelines and toerail are of real value at sea, where their adverse effect on visibility is a minor consideration. They should extend far enough forward of the cockpit to protect against spray blowing aft when you are beating to windward.

If you don't already have them, be sure to install nets under the lifelines forward to prevent lowered headsails from washing overboard. These nets will also provide some security for the crew working forward.

When you are sailing a long distance with a small crew, self-steering gear can be invaluable. The most reliable vane gears, which steer a course relative to the wind direction, are quite elaborate and must be permanently bolted to the stern, but there are simpler types without underwater parts that can be more easily removed if desired when the passage is over. Self-steering may also be accomplished in a relatively simple way by connecting the sheets to the helm. A method I sometimes use when reaching is to secure the jibsheet to the windward side of the tiller (or leeward side of a wheel) and balance the pull of the sheet with a piece of elastic shock cord attached to the opposite side of the helm. Another alternative is an electrically powered autopilot, which can either steer a compass course or be hooked up to a very small wind vane. Some autopilots are small portable units that cause minimal drain on the battery—as little as 2 to 4 watts, depending on the size and balance of the boat.

When making a passage that involves considerable night sailing, you need to be continually aware of the condition of your batteries. There should be at least two heavy-duty batteries, one for the engine and the other for general services, such as lights and electronic gear. It is best to avoid using the engine battery for general service because the engine's starting motor draws considerable current. As mentioned in Chapter 2, I have a battery condition meter on *Kelpie* to monitor the state of charge of each battery. One of *Kelpie*'s batteries is charged with the engine generator, the other with an alternator. The latter provides a relatively quick charge, but not always a reliable one, as alternators can be prone to failure. The generator, on the other hand, provides a slower but more reliable charge. A switch allows either battery to be charged by either system, giving reasonable assurance that our batteries will not go dead during the night.

## CREW SELECTION

For a long passage involving watches at night be sure you have a dependable crew. The essence of dependability is being continually available, alert, able, and willing to perform in adverse conditions. Jobs that need careful attention at night are steering, sail handling, navigating, and keeping a lookout. Naturally, you don't want a crewmember who is incapacitated by seasickness or fatigue, and the time to find out about these problems is before the passage, not during it. Motion sickness remedies can be very helpful in alleviating seasickness, but remember that with most kinds there may be undesirable side effects, including extreme drowsiness.

Surprisingly, some youthful athletic types are highly susceptible to fatigue and are sometimes overwhelmed by sleepiness at night. I know several experienced ocean racing skippers who prefer older sailors to take over night watches because youthful crewmembers have repeatedly fallen asleep or been inattentive during the late night or early morning hours.

## WATCHES AND ROUTINE

Watch methods are based on the number of crew on board and the number needed on deck for competent operation of the vessel. Just how the watches are organized depends

*It is the safest policy to lower the spinnaker just before the sun goes down if you are not racing.*

to a large extent on whether or not the boat is racing. Competitive, full-crew racing generally requires ample hands on deck at all times, while offshore cruising may need only one or two crewmembers on deck unless there is an emergency. The most commonly used systems divide the total crew in half and rotate the watches so that the same crewmembers do not stand watches during the same hours each night. The popular "two-watch" or "watch-and-watch" system uses four-hour periods and alternates the watches by "dogging," that is, dividing one period in half, normally the one from 1600 to 2000. The so-called Swedish systems divide the day into five watch periods, thereby automatically alternating watches. One method uses watches of six hours during daylight and four hours at night, while another uses watches of four-, five-, and six-hour periods with the shorter watches at night. For a one-night race or a cruise on which a number of crewmembers want to be (or are needed) on deck, a continuous rotating watch system is sometimes used. In this case one person goes off watch and is replaced by a new person coming on watch every hour.

On shorthanded passages, I think it is often desirable not to alternate watch times, but instead, to have the same crewmembers stand watches during the same hours every night. Individuals have different lifestyles and metabolisms. Some people tend to fall asleep early in the evening but have little trouble staying awake in the early morning; others like to stay up late but are usually groggy the next morning. These factors should be considered when making up the watch schedule, as well as the duties of certain crewmembers such as the navigator, who will need to be on watch at dusk and just before dawn to take sights. On some vessels with plenty of crew, navigators and cooks are exempt from standing watches.

Singlehanders must sleep, obviously, and this normally means that for nearly a third of the day the boat will have no lookout. During these periods, at night especially, there is risk of collision. In crowded waters the singlehander should make every effort to stay awake and stand watch, limiting his sleeping to short naps. When sleeping at night, he should show additional Rule 27 lights: two red lights, one above the other, indicating the vessel is not under command.

Activities of the on-watch crew include navigation, sail handling, and steering, all of which have already been dealt with. Another important duty is standing lookout. In crowded waters, of course, there should be a continuous watch for other boats, and on fully crewed vessels at least one person other than the helmsman should be assigned this duty. When offshore away from steamer lanes, the on-deck crew need not be so concerned, but even then, someone should make a complete visual sweep of the horizon at least every ten minutes. In remote waters ships have a way of sneaking up on you, and even seasoned sailors can be surprised by the speed at which some vessels move. Always sweep the horizon when the boat is on top of a wave, and be sure that all 360 degrees of the horizon are covered. When the skipper is off watch and a serious problem arises with navigation, light identification, ship proximity, or even sail changes, the skipper should be consulted. Remember John G. Alden's famous order: "Call me if it moderates."

When off watch, crewmembers should get all the sleep they can, as this will prevent fatigue and maintain judgment. It is a well-accepted fact that lack of sleep and fatigue can lead to hallucinations, or "tricks of the mind," that can cause serious errors in navigation and seamanship. Hallucinations are more common with singlehanders and doublehanders, but they can occur on any boat far offshore in stressful situations. When changing watches, the off-going crew should be sure that the on-coming crew is thoroughly awake and well-oriented before going below. The on-comer(s) must be informed of the compass course, condition of the weather, lights that can be seen and their identification, sail changes made during the previous watch, and any other pertinent matters. Be sure your berth has a lee cloth, bunk board, or safety belt so that there is no chance of your being rolled out in rough weather. Make every effort to keep your bedding and clothes dry.

## STORMY WEATHER AND FOG

There is no way of guaranteeing troublefree weather for a passage, but careful planning can certainly help avoid major storms and reduce the risk of fog. As mentioned earlier, a thorough study of the pilot charts and publications such as cruising guides and *Ocean Passages of the World* will help you select the optimum route and time of year. Good advice can also be obtained from the U.S. National Weather Service (or the equivalent in foreign countries) or private ship routing organizations such as Oceanroutes, Inc. (headquartered in Palo Alto, California), although the latter caters mostly to commercial shipping. Accurate weather forecasts cannot be made beyond five days, but current radio broadcasts can and should be obtained en route. Particularly valuable are the short-range continuous broadcasts on VHF/FM for coastal passages and, when far offshore, the hourly high seas storm information reports using the facilities of the time signal stations WWV and WWVH. There are also small weather facsimile instruments, such as the Alden Marinefax Recorders, that supply up-to-date synoptic charts.

Generally speaking, long offshore passages should not be made during the peak of the hurricane season, which, in the North Atlantic, is from about mid-August to mid-October, with the peak near mid-September. In the eastern North Pacific, hurricanes tend to come a little earlier, with the peak near the end of August, but the Pacific "hot spot" known as El Niño can throw off this timing considerably. In general, the safest time of year to venture offshore is in mid-June, after the unsettled spring weather but before the hurricane season, and indeed, many ocean races are held near this time. The passagemaker should become familiar with the tracks, signs, and other aspects of both tropical and extratropical storms. Modesty doesn't prevent me from recommending the book *Heavy Weather Guide* (Second edition, Naval Institute Press, 1984) which I had the good fortune to coauthor with the distinguished meteorologist Admiral (Ret.) William J. Kotsch. During a passage, it is important to keep track of the positions and movements of depressions (lows), high-pressure systems, and weather fronts, not only with the radio or facsimile equipment (if you're fortunate enough to have it), but also by constantly observing the sky while monitoring wind direction and atmospheric pressure with a barometer.

Thunderstorms have been discussed in the last chapter, but one further point might be made about timing. Close to shore, thunderstorm clouds (cumulonimbus) of the isolated air mass type tend to develop in the late afternoon, and squalls will often occur before dark in the right conditions of heat and humidity. Offshore, however, squalls may be more likely after dark, sometimes as a result of lower, cool air being heated by a warm sea and pushing up unstable clouds.

The cruising sailor is most often affected by two types of fog, advection and radiation fog. The former is probably more frequently encountered, particularly by the offshore passagemaker. Fog occurs when water vapor in the air condenses and forms into suspended droplets. Condensation may take place as a result of air saturation from increased moisture or from the air cooling to its dew point, the temperature at which condensation takes place. Advection fog forms when warm moist air is cooled as it moves over cold water, as is common, for example, off the coast of Maine when southerly winds blow over the cold water carried south by the Labrador Current. This happens as readily at night as during the day. Weather reports can warn of advection fog not only directly, but often indirectly by predicting temperature and wind direction. Some reports also give the dew point, and when this is close to the actual temperature of the air, fog is probable. On the other hand, a large spread between the temperature and dew point indicates that fog is unlikely. Some boats that cruise extensively in foggy waters carry a sling psychrometer, an instrument that measures relative humidity and dew point.

Radiation fog, also known as ground fog, is most common at night, but it rarely extends very far offshore. It tends to form in relatively calm, clear weather when the land cools, and humid air just above the ground is cooled to its dew point. This fog will be dissipated by a fresh wind, but even in calm weather it will usually burn off a few hours after sunrise.

In planning your route, it generally pays to head offshore to avoid radiation fog, but it may be beneficial to head closer to shore into warmer water to avoid the thickest advection sea fog. When selecting an anchorage, try to pick one far up an open bay, where the water will likely have been warmed by the surrounding shore and is thus less apt to cool the humid air to its dew point. It is risky to hug a shore too closely during a foggy night. Not only is there the possibility of grounding, but also of encountering hard-to-see objects, such as small buoys or fish traps. When possible, try to sail at a moderate speed rather than using the engine, so that you can hear sounds more easily and reduce the risk of tangling your propeller in fish lines or nets. Lay out your courses away from fish trap areas (as shown on the chart) and between prominent navigation aids equipped with sound signals. Of course, Loran, RDF, and radar are invaluable for low-visibility navigation. Be sure to learn the fog signals that vessels must sound when underway or anchored. They are given in Rule 35 of the *Navigation Rules*. A more extensive discussion on this subject can be found in a companion book in the International Marine Seamanship Series, *Sailing in the Fog* by Roger Duncan.

A night passage can be a rewarding experience that will be remembered vividly for many years. Taking the helm after dark increases your awareness by bringing all of your senses into full play. Even if you are not steering, but merely relaxing between navigational checks or looking out, you have a rare opportunity to cogitate and commune with the sea. In my opinion, sailing at night is a necessary ingredient for complete fulfillment of the yachting life, and it adds yet another dimension to the joy of messing about in boats. May every sailor accept the challenge and reap the pleasures.

# Index